PREFACE

After discussion of OECD's first biotechnology report (*Biotechnology: International Trends and Perspectives*, Paris, 1982), the Committee for Scientific and Technological Policy agreed to concentrate its future efforts on four issues: patents, safety, government policies and long-term impacts. The Committee also recommended that work on long-term economic impacts be undertaken only after the other three activities had been completed, or at least had made substantial progress.

Thus, a first expert workshop, sponsored by the Italian authorities, to review economic impact issues took place only in 1985 at Castel Gandolfo, and industry interviews to examine current industrial strategies in biotechnology, began in 1986 and were completed during the second half of 1987.

A first draft of the present report was discussed during an OECD Expert Seminar on Long-Term Economic Impacts of Biotechnology, sponsored by the Government of the United Kingdom on 4th-5th February, 1988 in Admiralty House, London, as well as by a joint session with United Kingdom officials which took place during this Seminar (List of Participants in Annex). The OECD draft was extensively revised after this Seminar.

During an earlier discussion of this project, the Committee wished to see its scope restricted in view of the enormous difficulty of the subject, but also wished to have the social impacts included. The following Introduction will indicate why it has been so difficult to define and assess "social" impacts. The experts who helped the Secretariat discussed this problem on various occasions, and proposed to abandon the term "social impacts" in the title of this report and to replace it by "wider impacts".

Many, if not most, of the trends and questions mentioned in the report, concern the next ten years. The Chapters on diffusion and impacts (III, VI, VII), however, involve longer time perspectives as well. In spite of this, it was considered to be more appropriate to delete the word "long term" from the original title of the project, and to speak simply of "economic impacts".

In order to support the economic analysis, particular efforts were made to gather economic and R&D statistics which should have provided a new, original data base. A micro-economic survey of industrial company policies, by Dr. R. Hoeren from Prognos AG., Basle, which involved 94 companies in 17 OECD countries and led to a background report[72] was more difficult than had appeared during the first trial interviews. During the last two years, with competition and commercial interests in biotechnology increasing quickly, industry has become more reluctant to reveal facts and figures than had first been hoped. A parallel effort to gather national data on R&D expenditures in biotechnology, particularly on industrial expenditures, again met with all the difficulties which have so far prevented the establishment of a comparative international R&D data base on biotechnology. Definition is still the most intractable, but not the only difficulty.

Nevertheless, a beginning was made and first data indicate that the importance of biotechnology in industrial R&D expenditures is increasing. It is hoped that more comprehensive and comparable data can be published in the coming years.

As public attitudes and occasionally also official policies towards biotechnology have tended to swing between euphoria and scepticism, which are both unjustified, the need for reliable R&D and economic data on this emerging field remains urgent, at national and international levels.

The report consists of three main parts. The first (Chapters I, II) speaks of the development and *production* of biotechnology, looking into scientific and technical trends for the next decade or more and into current industrial policies and problems; the second (Chapters III, IV, V) reviews the *diffusion* of biotechnology in the medium and long term and the related structural adjustment questions, the third part (Chapters VI, VII) focusses on some economic *impacts* of biotechnology. As quantitative, macro-economic forecasts of such impacts are still impossible, an effort was made to provide an analytic framework which should allow governments to assess the most probable directions of biotechnology impacts on trade and competitiveness, and on employment.

The question of *definition* has been as difficult in this as in earlier reports. These are the principles which have been followed, and the terms which have been used:

Biotechnology, in this report sometimes called *Classical Biotechnology (CBT)*, is thousands of years old. During the last several decades, numerous scientific and technological advances have turned biotechnology into an increasingly efficient set of techniques, called *Modern Biotechnology*. Since the late 1970s, the discovery, particularly of recombinant DNA-techniques and of cell-fusion, has led to a radical acceleration of progress and to a multiplication of both tools and applications. This is called *New Biotechnology (NBT)*.

Biotechnology, and particularly new biotechnology is not a distinct sector, but a *broad enabling technology* which will probaly affect many sectors of the economy.

It is not in contradiction with this that the report also uses the terms *Biotechnology industry* or *Biotechnology companies*. These refer to the industrial enterprises which carry out the bulk of industrial R&D in modern and new biotechnology, irrespective of the sector of economic activity. These terms can occasionally also include traditional companies which use or manufacture classical biotechnologies.

While considering biotechnology as a continuum, the report focusses primarily on the production, diffusion and impacts of *new biotechnology*. Where necessary, classical or modern biotechnology are mentioned and included in the analysis (Chapters II, VI).

Whenever a precise definition of biotechnology was required, particularly during the interviews with industrial companies, the OECD definition of 1982 (*Biotechnology: International Trends and Perspectives*, p. 21) was used: "Biotechnology is the application of scientific and engineering principles to the processing of materials by biological agents to provide goods and services."

Although the report's arguments about economic aspects and impacts are based on this definition, some reflections go beyond it and refer to possible *human applications* of the tools and discoveries which are the basis of all new biotechnologies (Introduction, Chapter I).

The difficulties of definition, and the resulting misunderstandings, explain recent suggestions to abandon the term "biotechnology" as too general, and to replace it by the precise term of whatever specific technology or application may be discussed (Chapter IV,

BIO TECHNOLOGY

ECONOMIC AND WIDER IMPACTS

ORGANISATION FOR ECONOMIC CO-OPERATION AND DEVELOPMENT

Pursuant to article 1 of the Convention signed in Paris on 14th December 1960, and which came into force on 30th September 1961, the Organisation for Economic Co-operation and Development (OECD) shall promote policies designed:

- to achieve the highest sustainable economic growth and employment and a rising standard of living in Member countries, while maintaining financial stability, and thus to contribute to the development of the world economy;
- to contribute to sound economic expansion in Member as well as non-member countries in the process of economic development; and
- to contribute to the expansion of world trade on a multilateral, non-discriminatory basis in accordance with international obligations.

The original Member countries of the OECD are Austria, Belgium, Canada, Denmark, France, the Federal Republic of Germany, Greece, Iceland, Ireland, Italy, Luxembourg, the Netherlands, Norway, Portugal, Spain, Sweden, Switzerland, Turkey, the United Kingdom and the United States. The following countries acceded subsequently through accession at the dates indicated hereafter: Japan (28th April 1964), Finland (28th January 1969), Australia (7th June 1971) and New Zealand (29th May 1973).

The Socialist Federal Republic of Yugoslavia takes part in some of the work of the OECD (agreement of 28th October 1961).

Publié en français sous le titre:

BIOTECHNOLOGIE
EFFETS ÉCONOMIQUES
ET AUTRES RÉPERCUSSIONS

2*c*). Although some countries will not find it easy to accept this proposal, there is value to it, and it deserves to be discussed. It is the growing pervasiveness of the biotechnologies which might ultimately lead to this replacement.

The authors who drafted this report, experts as well as Secretariat members, did so in close co-operation. They are F. Chesnais (Chapter VI), C. Freeman (Chapter III), R. Galli (Chapter V), E. Jaworsky (Chapter I), and S. Wald (Introduction, Chapters II, IV and VII). The chapters were integrated and edited by S. Wald of the Secretariat.

Thanks are due to the Government of the United Kingdom, and particularly Dr. Ron Coleman, Chief Scientist, Department of Trade and Industry, and Dr. Roy Dietz, Deputy Director, Laboratory of the Government Chemist, and his staff for sponsoring and organising the OECD Expert Seminar in London which made a thorough review of this report possible. Thanks are also due to the Commission of the European Communities (Directorate General for Science, Research and Development), and the Government of the Federal Republic of Germany (Federal Ministry for Research and Technology), who helped finance the industry survey through generous grants. Finally, thanks should also be expressed to the Italian authorities, and particularly to Professor Umberto Colombo, President of ENEA, Rome, who co-sponsored and chaired the first OECD Workshop on this subject in Castel Gandolfo in 1985.

This report was reviewed by the OECD's Committee for Scientific and Technological Policy at its 50th Session in October 1988. However, it does not necessarily reflect the views of the OECD or of its Member governments.

Also Available

BIOTECHNOLOGY AND THE CHANGING ROLE OF GOVERNMENT (1988)
(93 88 04 1) ISBN 92-64-13072-1 148 pages £11.00 US$20.00 FF90.00 DM39.00

SCIENCE AND TECHNOLOGY POLICY OUTLOOK (1988)
(93 88 03 1) ISBN 92-64-13076-4 122 pages £9.50 US$18.00 FF80.00 DM35.00

EVALUATION OF RESEARCH. A Selection of Current Practices (1987)
(92 87 05 1) ISBN 92-64-12981-2 78 pages £5.00 US$11.00 FF50.00 DM22.00

RECOMBINANT DNA SAFETY CONSIDERATIONS. Safety Considerations for Industrial, Agricultural and Environmental Applications of Organisms Derived by Recombinant DNA Techniques (1986)
(93 86 02 1) ISBN 92-64-12857-3 70 pages £6.00 US$12.00 FF60.00 DM27.00

BIOTECHNOLOGY AND PATENT PROTECTION. An International Review by F.K. Beier, R.S. Crespi and J. Straus (1985)
(93 85 05 1) ISBN 92-64-12757-7 134 pages £8.00 US$16.00 FF80.00 DM35.00

BIOTECHNOLOGY. International Trends and Perspectives by Alan T. Bull, Geoffrey Holt, Malcolm D. Lilly (1982)
(93 82 01 1) ISBN 92-64-12362-8 84 pages £5.50 US$11.00 FF55.00 DM28.00

* * *

STI REVIEW – SCIENCE, TECHNOLOGY, INDUSTRY – ISSN 1010-5247
(Half-yearly)

No. 4 December 1988
(90 88 01 1) ISBN 92-64-13172-8 190 pages

Per issue: £12.90 US$24.00 FF110.00 DM47.00

1989 Subscription £21.00 US$40.00 FF180.00 DM78.00

Prices charged at the OECD Bookshop.

THE OECD CATALOGUE OF PUBLICATIONS and supplements will be sent free of charge on request addressed either to OECD Publications Service, 2, rue André-Pascal, 75775 PARIS CEDEX 16, or to the OECD Distributor in your country.

TABLE OF CONTENTS

LIST OF TABLES

LIST OF FIGURES

EXECUTIVE SUMMARY

Biotechnology is not a sector, but a broad generic technology. It had a slow, empirical development during centuries. Modern R&D, and particularly the recombinant DNA, cell-fusion and other breakthroughs of the last decade, have turned it into an efficient and rapidly growing set of tools and applications, a "new biotechnology" with vast implications.

The report "Economic and Wider Impacts of Biotechnology" focusses on this new biotechnology, but takes earlier developments into account when necessary. It touches upon many issues which have social as well as economic relevance.

New biotechnology is distinguished from other major technologies of the 20th century by the fact that its impacts on the quality of life, its human and social consequences, are arriving earlier and may go deeper than macro-economic impacts measured by productivity, investment or GDP growth. Many discoveries in the underlying research fields provide new keys to the understanding of life and health, which will affect man and social relations more directly than discoveries, e.g. in materials or information technologies. The analysis and measurement of quality-of-life changes, however, is extremely difficult because service sector statistics and social indicators which should be able to measure some of them, have not made sufficient progress during the last years. Estimating the quantitative economic impacts of biotechnology is beset by other problems, for example by large variations in the definition of biotechnology, by reluctance of industry to reveal facts and figures in a sector where commercial interests and competition are increasing rapidly, and by the unpredictability of numerous important technical developments.

The report consists of three parts. The first, (Chapters I and II), speaks of the development of biotechnology, the second, (Chapters III, IV and V) reviews the diffusion of biotechnology and the related structural adjustments, the third (Chapters VI and VII) analyses some prospective economic impacts.

Chapter I summarises the major trends in science and technology, looking forward to the end of this century, if not beyond. Progress has been rapid during the last five years and will continue in many directions, with significant interactions and spin-offs occurring between various sectors (between human and animal health; between plant, animal and human diagnostics; between health and agricultural applications; between agricultural, food and environmental needs and opportunities, and others). Often, rapid advances lead to new technical bottlenecks or reveal fundamental research gaps which require a steady, if not growing, R&D effort.

Among the many encouraging recent research developments is the successful exploration of various microbes as alternative hosts to *E. Coli* for gene expression, the remarkable progress in plant genetic engineering and cell and tissue cultures and the progress in isolation and purification techniques of biological macro-molecules.

In the *pharmaceutical* sector, the rate of introduction of new products has clearly increased since 1986, including that of natural proteins and peptides which have been turned into useful drugs (insulin, human growth hormone, interleukins, growth factors). The main emphasis, however, is on diagnosis, prevention and immunology. While the AIDS problem is mobilising increasing biotechnological research attention, which hopefully, will lead to a solution before the end of the century, progress in gene therapy has not been rapid and will not lead to large-scale applications in this century.

In *agriculture*, the stakes of biotechnology are equally great, and research directions equally numerous. Biotechnology could increase and improve food production by increasing the growth rates of plants and animals, by creating plants resistant to diseases, micro-organisms, insects, herbicides, salt or harsh climates, by creating transgenic animals, by preventing and controlling animal diseases, to mention only some of the ongoing R&D projects. The momentum is accelerating in plant genetic engineering, where breakthroughs can soon be expected in the capability to transform cereal grains (maize, wheat, rice).

Biotechnology will make numerous contributions in the *food and feed* sector and in the *chemical* sector although with present oil prices, no significant impact should be expected on commodity chemicals. The development of *environmental* uses of biotechnology (pollution control, waste treatment) may be slow during the next decade.

Apart from the main areas of possible future application, several present and future key areas of research and generic technology are also reviewed: biosensors; human genome research; organic computers and biochips, where revolutionary concepts are being probed, but where also totally new synthetic approaches will be necessary to make real progress.

This large number of promising scientific and technical developments indicates a potentially very broad range of applications for new biotechnology.

However, the actual range is much narrower than the potential range. Chapter II, based on interviews with 94 industrial companies which are active in various sectors of application for biotechnology in 17 0ECD countries, touches upon some of the reasons.

The early years of industrial R&D in biotechnology have been dominated by a strong "science-push". Product development often focussed on scientific and technological feasibility rather than on real market needs, and not enough attention was paid to costs. During the last two or three years, companies have become more concerned with technical limits, costs and market demand.

The overwhelming majority of the 94 interviewed companies have plans (which does not always mean budgets) in biotechnology, and approximately two-thirds intend to develop new (i.e. rDNA-derived) biotechnologies. Most of them aim at new products, some of them at new processes. Diversification into new biotechnology is very noticeable in the pharmaceutical and, albeit to a lesser extent, in the agro-industrial companies; it is weak in the food and feed sector. Considering that many of the 94 companies are small and/or not in particularly "science-based" sectors of the economy, the high overall proportion of companies moving into biotechnology seems to indicate that this movement is both broadly based and fast.

In fact, many companies, particularly those which aim at large future markets, see in biotechnology a new generic technology requiring a pool of permanent in-house know-how. Hence, R&D strategies are the main component of current industrial strategies in biotechnology, or at least the component about which information has been made available, whereas none is provided on the identification of possible new products. Many companies plan to increase their biotechnology R&D expenditures during the next years, both in relative and absolute terms.

The large companies often devote an important part of their R&D policy to strategies of co-operation with small, specialised biotechnology start-up companies. This gives them greater flexibility and reduces overheads. The ability of those small companies to implement fast changes is one of their main advantages. The more traditional small and medium-sized companies, however, face large disadvantages in biotechnology R&D, at least in Europe.

While the number of new products presently on the market is relatively small and includes some which may be controversial such as bovine growth hormone, there are interesting market potentials for the future.

Current policies of industrial companies have implications among others for employment, industrial concentration and international trade. Due to cost-reduction policies, overall employment in biotechnology companies is unlikely to increase in the coming ten years although employment in R&D will increase. High R&D and marketing costs for new biotechnology products will continue to promote co-operation strategies between companies, but will also favour the big rather than the small establishments. The industrial structure which is emerging is one of "decentralised concentration", with numerous small companies depending upon relatively few big companies, but contributing essential research and innovations to their partners.

New biotechnology is clearly a technology of highly industrialised countries, both with regard to R&D requirements and market potential. Companies will exploit the advances in plant genetics to replace Third World crops, which might be increasingly grown in OECD countries, thus reinforcing the concentration of world trade within the OECD area.

Most of the interviewed companies have been critical with regard to their government's support policies in biotechnology, although these complaints are very heterogeneous.

From the short-term perspective of Chapter II, the focus of the report moves back to very long-term considerations.

Chapter III discusses the future diffusion of biotechnology through the economy and particularly the probable time-scales. Parallels are drawn with the diffusion of the electronic computer and of electric power. It is underlined that it is very important to recognise the differences as well as the similarities between the various generic technologies which have so deeply transformed industrial societies over the last hundred years.

The most important lesson which can be learnt from the introduction and diffusion of revolutionary new technologies is that the changes in capital stock, skill profile, industrial structure and social organisation which they require are a matter of decades rather than years. Recognition of the relatively long time-scales involved, helps to avoid errors both of technological optimism, which tends to ignore some hard economic realities, and of conservatism which fails to recognise the enormous long-term potential of pervasive technologies. Both errors can be found in the history of the electronic computer, where some of the involved experts were completely unable to see the great potential, whereas others believed in the imminence of large-scale economic and social consequences twenty years before they began. Similarly, it took almost half a century between the first appearance of electric power in the 1880s and the large-scale replacement of steam by electricity as the main source of industrial power.

There is, in fact, a major difference between the diffusion process for a single, even radically different product which can be very fast, and the diffusion process of a generic technology with numerous applications in a vast number of economic sectors, leading to a new "technological paradigm" and a new technological common sense for a generation of engineers, managers and consumers.

To achieve a major change of a technological paradigm, five conditions should be fulfilled; *i)* a new range of technically improved products and processes; *ii)* cost reductions for many of these; *iii)* social and political acceptability; *iv)* environmental acceptability, and *v)* pervasive effects throughout the economic system.

Computer-based information technology clearly satisfies all five criteria. Nuclear power does not. The case of new biotechnology is not yet clear-cut. It satisfies the first, but not yet fully the second criterion. Comparative costs of biotechnology production remain high in certain sectors, but may come down as the number of profitable innovations grows. Also, it was inevitable that biotechnology encountered problems of social and environmental acceptance. Some of these problems may remain, although much progress towards acceptance of biotechnology has already been made. Regarding the fifth condition, biotechnology is more pervasive than many other technologies as it has already found applications in agriculture, industry and the service sectors, but it is at present less pervasive than information technology which has been able to penetrate almost all products and processes of human activity.

In this century, biotechnology will not become a predominant technology for most industries and services. The time scale for biotechnology to become a major base for GDP growth and investment is not likely to be much shorter than the time scale of earlier pervasive technologies which also depended upon far-reaching structural adjustments. This means that biotechnology may begin to have major macro-economic effects from the second decade of the next century on. In the meantime, its economic importance for individual sectors, for profits and competitiveness, is likely to grow steadily, and its qualitative implications, through impacts on life and health, will become a major challenge.

Among the conditions which promote the diffusion of new technology, two are particularly important for biotechnology; public acceptance and patent protection. They are discussed in Chapter IV.

Biotechnology differs from nuclear power technology, where public acceptance has been difficult, in three ways: it includes a wider and more diverse range of technical tools and approaches (microbes, plants, animals and parts thereof), its health and environment impacts have more positive than negative connotations, and most importantly, the discussions of risks and benefits have begun earlier than for any other technology of this century and before the first industrial investments have been made, and this has already shaped the rate and direction of technological development.

Scientists are now more optimistic regarding safety than they had been 15 years ago, and differences between experts are more rare and narrower than at any time before. A public and institutional framework to review biotechnology safety has been put in place in most OECD countries, and a movement towards international harmonisation of rDNA criteria has begun. However, in contrast to the scientific assessment of rDNA safety, there has been little or no convergence of public attitudes towards biotechnology. These can vary widely, depending upon cultural traditions related to food, health, medicine etc.

Governments, parliaments and some industrial companies have paid attention to the public acceptance question, and have developed a wide range of responses. Whilst it is not difficult to understand what the public's main concerns are, it is more difficult to decide who should be addressed – the public in general, politicians and journalists, or other specific target groups.

There can be no unified response to public concerns, which may be influenced by various unpredictable developments in the future. However, it can be expected that an increasing number of useful products arriving on the health and environment markets will influence the public discussion in the coming years more in favour of biotechnology.

Turning to *patent protection*, it is underlined that modern biotechnology has depended from its beginning upon an appropriate legal framework. However, in no other field of technology do national laws vary on so many points and diverge so widely as they do in biotechnology. Although the decision of the US Supreme Court in 1980 to accept the patenting of man-made micro-organisms was a great legal breakthrough which has been followed by many other countries, several other patent problems have not yet found a solution satisfactory to the inventor in biotechnology.

They include the question of a "grace period" allowing scientists to submit a patent claim on an invention even if they have already disclosed it in scientific publications, which is accepted by some, but not by other (i.e. European) national laws; the extention of patent protection to genetically modified plants and animals which the European Patent Convention (1973) prohibits, and the insufficient length of patent protection. Better international harmonisation of patent protection is essential for the large-scale diffusion of biotechnology.

Diffusion will also have to be accompanied by considerable structural adjustments in the sectors affected by biotechnology. Chapter V reviews this need in two sectors, public health and agriculture.

Economic, social and institutional adjustment needs results from specific trends in technical change which biotechnology will accentuate. Among these trends are a growing emphasis on diagnosis and prevention, "dematerialisation" or the replacement of heavy by light, and of expensive by cheap or abundant materials and other production factors, and a growing rationalisation of the innovation process.

In the *health* sector, the new diagnostic tests will lead to more far-reaching change than new therapeutic approaches, although in the longer term, these will also have important consequences. At present, the fastest growing biotechnology market is that for diagnostics (monoclonal antibodies, biosensors, gene probes) which can be applied to humans, animals, plants, the environment and industry. The rapidity, specificity and facility of use of the new tests, and their wide spectrum of applications, heralds an era of mass diagnosis extending to the general population and reducing the damage caused by disease.

To be effective, biological innovation has to be accompanied by innovation in other areas such as automation of testing procedures and of instrumentation, and management of systems for the processing, interpretation, transmission and retrieval of large numbers of analytic data. Also, the development of many simple diagnostic kits will favour a transfer of tests from the laboratory to the doctor or to the private individual (home tests), which represents a major functional innovation. However, for diseases such as AIDS or cancer, home tests could be problematic.

The shift towards diagnosis will, among other adjustments, require retraining of doctors and a modification of reimbursement schemes and other health regulations. Since this breaks with traditional practice, obstacles from groups and administrations which have vested interests in the traditional system, can be predicted.

The cost implications of a move towards diagnosis and prevention are complex and controversial. The hope has been expressed that this will help reduce public health budgets, but this is not certain.

The new trends will also have important effects in the pharmaceutical industry which is already changing from a supplier of products to a health care industry, relying on a wide, interdisciplinary R&D base. The critical success factor in this new industrial challenge is the capacity to run complex systems rather than biotechnological knowledge in itself; hence, the electronics industry might find itself at an advantage in relation to the pharmaceutical industry. In countries where the public health sector has found it difficult to adapt to technical change, governments might even turn to private enterprises to extend their activities to diagnostic and other public health services.

Impacts on *agriculture* could be equally profound. So far, these have largely been quantity impacts. The powerful tools of new biotechnology have first been directed towards increasing production or decreasing production costs; through improvements in plants and animals, reduction of disease etc. The introduction of new biotechnological know-how on a broad front might bring about a further increase in agricultural production, exacerbating some of the current surplus problems.

Thus, it is evident that agricultural biotechnology should be directed more towards qualitative than quantitative goals. This includes food with better taste and aroma, safer food, food with fewer chemical residues which may imply the replacement of present agro-chemicals by biological techniques, and greater specialisation and diversification of food products in order to respond to specific demands.

Moreover, the development of new and economically viable, particularly industrial utilisations of agricultural products has become a critical challenge. By contributing to this goal, biotechnology could become a decisive tool in the necessary transformation of the agriculture of OECD countries. For example, the growing need for biodegradability and environmental compatibility might be better satisfied through products based on transformation of biomass than through synthetic products.

Shifting the emphasis from quantity to quality will call for growing interaction and convergence between agriculture, industry and services, and a transformation of agribusiness. In fact, diversification and development of new crops can succeed only with extensive co-ordination between agriculture and users. Simultaneously, industry offers agriculture an ever growing range of inputs and services. While industry may play the most dynamic role in this transformation, it is governments and the farmer lobbies which will exercise most influence on the speed and direction of change.

Chapters VI and VII now turn to some of the prospective long-term economic impacts of biotechnology, impacts on trade and competitiveness, and on employment. As no quantitative forecasts are possible, the report endeavours to establish an analytic framework which might help governments to evaluate current trends and to put them into a longer time perspective.

Chapter VI recalls that the enhancement of industrial competitiveness has, up to now, been the overriding goal of government policies to support biotechnology. The pursuit of such a goal requires an international perspective on the potential trade and competitiveness impacts of biotechnology, which in view of the limited number of new products and the absence of relevant trade statistics, can only be based on economic theory and historical experience with modern biotechnological products which were traded already before the development of the most recent rDNA-derived products.

Since the Second World War, resource endowments leading to comparative advantages have been based less and less on climatic and geographical factors, and more and more on R&D, innovation and investment, which explains at least partly, the decline of the share of developing countries in world merchandise trade since the early 1970s and the increasing concentration of world trade in the OECD area.

Innovative capacity can increasingly be viewed as the basic source of difference in comparative advantage between countries and as a "chronic disturber" of existing patterns of trade. Countries and industries with strong R&D and innovative capacities are likely to be at the initiating end of trade-modifying processes, or they are likely to be able to adapt to them successfully, and weak countries and companies will often be at the receiving end of such processes and will have to bear the full brunt of painful employment and income adjustments.

At the moment, competition in biotechnology is waged principally among OECD countries, while most trade impacts are experienced by some developing countries. In the future, genetic modifications of cereal grains (wheat, maize, rice) could lead to trade impacts for cereal-exporting OECD countries as well.

In the longer term, biotechnology will have both strong trade-creating and trade-displacing effects. The potentially wide range of biotechnology might offer unique opportunities for creating totally new products and large markets for which there is presently no competition and no issue of competitiveness. Thus, biotechnology in a broad sense, and policies to support biotechnology, might lead to fewer trade tensions than technologies where competitors offer almost identical products in slow-growing markets (aircraft, automobiles).

For OECD countries, being "competitive" in biotechnology presently still means essentially preparing for the future, by reinforcing the scientific, technical, manpower, industrial, data-bank and other components of the biotechnology infrastructure.

While these potential trade-creating effects are not yet visible, the trade substitution effects are, because they have their origins in the advances of biotechnology even before the advent of rDNA techniques.

Biotechnology is reducing the overall demand for primary products from developing countries. A review of currently predictable trade impacts of biotechnology in agriculture points out that this reduction is a continuation of an old pattern whereby industry was capturing markets away from agriculture. Three biotechnology examples are mentioned: enzyme-based sweeteners which have reduced sugar production and exports particularly from developing countries; single-cell proteins although these can presently not compete with agricultural proteins as animal feeds (soya), and *in vitro* plant propagation and cell tissue culture which will become an enormous source of competition and substitution for potentially all crops because these techniques might be applied to all plants, in other words, in a much wider way than synthetic chemical products which have already replaced many Third World crops in the past (indigo, jute, rubber).

Already, the cloning *in vitro* of oil palm trees which have so far grown only in tropical developing countries could lead to increased competition between different raw materials used for oil and fats.

The Chapter also looks at prospective trade impacts in the pharmaceutical sector. As this sector is strongly shaped by direct foreign investment (multinationals), only a small proportion of pharmaceutical output is traded across national borders. Thus, biotechnology will affect and diversify pharmaceutical production inside countries many years before any significant trade impacts will emerge.

Finally, Chapter VII reviews prospective employment impacts of biotechnology. Again, the absence of statistics on employment in biotechnology-related companies or activities makes quantitative analysis nearly impossible. But, in view of the size of the sectors which will be affected by biotechnology, the question of employment impacts is a legitimate one. These sectors are, agriculture, public health and the food, chemical and pharmaceutical industries, which employ more than 10 per cent of all civilian employees in the most industrialised OECD countries, and up to 20 per cent or more in countries with larger agricultural sectors.

Past experience shows that global employment levels in the OECD area have been much less influenced by technology than by macro-economic factors, and that technology-induced job losses (mainly in manufacturing) have been at least compensated by technology-induced job gains (mainly in services).

A discussion of prospective employment impacts of biotechnology should distinguish between four categories: direct effects on suppliers (biotechnology companies), direct effects on users (agriculture, health, industry), indirect effects through investment multipliers, and indirect effects through higher demand resulting from higher income or lower costs.

The job effects in the biotechnology industry will, in the short and medium term, be dominated by rationalisation and cost-cutting measures particularly of the large corporations.

In the longer term, employment could partly depend upon trends in innovation; new products are likely to create more new jobs than new processes or substitution products, innovation is likely to create more new jobs when it comes from small rather than large companies, and the factor bias (labour versus capital) in innovation could also play a role; there might be limits to labour-saving automation in biotechnology due to the complexity inherent to some of the new processes or products.

Certainly, qualitative employment impacts will be important, and their direction is easier to predict than that of the quantitative impacts. A higher qualification profile will be a predominant feature of the biotechnology industry of the future, requiring a continuous retraining effort.

Among the numerous employment effects in the user sectors, those expected in agriculture are certainly the most critical ones. They are the only ones which have already provoked counter-measures. Thus, the diffusion of some innovations (enzyme-based sweeteners, bovine growth hormones) has been delayed or held up, particularly in Europe, because of a perceived threat to agricultural jobs.

However, many OECD countries have their biggest agricultural adjustments already behind them and they should be less vulnerable to negative job impacts of biotechnology. Agricultural adjustments will continue, and employment will go down further even without biotechnology, because of the synergistic effect of so many other innovations on productivity. This is particularly true of Mediterranean and Third World countries with relatively large agricultural sectors. The prohibition of individual biotechnology products will not stop this historical process. Biotechnology, on the contrary, could facilitate agricultural adjustment if it were to focus more on quality improvements and on the development of industrial uses for agricultural crops.

Regarding the indirect employment effects, little is known about investment multiplier effects, except that biotechnology investments are growing.

A perspective on indirect effects through higher demand and cost reductions, is possible. In fact, possible negative, short-term employment effects must not be seen in isolation from other potential factor-saving effects. If labour-saving tendencies in biotechnology were matched by equivalent capital and other factor-saving tendencies, the net effect on employment could be positive in the long term, through reduced costs and/or increased global demand. While the costs of new biotechnology processes are still too high to be competitive in the bulk chemical and energy sectors, genetic engineering has opened opportunities for factor savings (not only in labour, but in capital, materials, energy) in other sectors, e.g. in the production of rare pharmaceutical substances.

The net result of so many possible trends cannot be precisely evaluated. In the short and medium term, biotechnology might add somewhat to the unemployment problem, particularly in countries with large agricultural sectors, although it will be impossible to separate biotechnology from other job-reducing effects of technology. In the long term, after fundamental advances in many, including health-related, sectors and with an increasing number of new products, biotechnology could well become a net creator of jobs. Monoclonal antibodies and genetic fingerprinting have already shown that this technology has a great potential for unexpected discoveries and products.

INTRODUCTION: BIOTECHNOLOGY AND THE QUALITY OF LIFE

1. New keys to understanding life

New biotechnology is distinguished from all other major technologies of the 20th century by the fact that its impacts on the quality of life – its human and social consequences, are arriving earlier and may go deeper than its economic impacts. It is one of the paradoxes of this technology that these qualitative impacts, which clearly have already begun, are difficult to analyse and measure, whereas the main macro-economic impacts which in principle, will be measureable, still lie many years ahead (Chapter III). Many discoveries of biotechnology affect human life and social relations more immediately and profoundly than the discoveries say, in materials or in information technologies, which explains why biotechnology has met with so much public interest.

The importance of qualitative or "social" impacts of biotechnology has been recognised very early, and a few attempts have been made to review them[1]. However, some confusion persists in the use of the term "social". Many people use the term social impacts synonymously with quality-of-life impacts. In other instances, the term covers the large-scale structural changes particularly in public health systems and in agriculture which new biotechnology may bring about and which are discussed in Chapter V. Other issues which may be included among social impacts are the safety and environmental concerns which are so important to society, or the legal and ethical questions linked to human genetics, or public acceptance of the new technology (Chapter IV), or impacts on employment, skills and working conditions (Chapter VII).

Of course, many aspects of biotechnology, particularly those relevant to health, nutrition, the environment, or employment, have as much "social" (or better: societal) as economic relevance, and their separation into two distinct types of impacts is often not obvious.

For example, it is very significant that the fastest growing market for new biotechnologies is that for monoclonal antibodies, discovered in the early 1970s and first commercialised in 1980. The revolution in diagnostics through monoclonal antibodies and gene probes will be discussed from various angles (Chapters I, II, V, VI). They allow for fast and precise identification of micro-organisms and other biological agents and will lead to dramatic improvements in the quality of life, with human and social benefits incommensurably larger than the direct economic benefits.

It is true that similar comments could be made about earlier pharmaceutical discoveries, but the gap between the economic and the human and societal impacts are particularly striking in the case of monoclonal antibodies because of their immediate and wide applicability and their low costs.

However, the tools of new biotechnology, particularly when applied to human health and life, have a wider quality dimension than is shown by the example of individual products which improve quality of life. New biotechnology may overtake chemical synthesis and conventional genetics in power and efficiency, but its most decisive contribution is likely to be in knowledge, and its most distinctive feature is its intimate link with man and life. Already, the discoveries and new techniques of the last fifteen years have revolutionised the basis of biological, medical and pharmaceutical research, and their contributions to the human condition may ultimately challenge existing attitudes and patterns of behaviour, thus bringing about sweeping changes in the way we see ourselves and maybe in the way we live.

Chapter I mentions the research projects on the human genome, planned in various countries. They will be the biggest projects ever undertaken in the life sciences, and are likely to last 20 or more years.

They include gene-mapping which means providing a complete topographic map of human genes as a basis for all future research on life, and gene sequencing which aims at understanding the exact nucleotic sequence of individual genes for living organisms as a basis for future commercial or, in the case of man, future therapeutic applications. The scientific, medical and ethical, but also the financial and societal stakes in the results are enormous.

Already today, early results of human genetic research may have profound quality-of-life impacts, for example, the invention of genetic or DNA-fingerprinting. This is based on the discovery that blood, other body fluids and hair contain fragments of human DNA which are specific to every individual. This allows for positive proof whether or not a blood sample (including several-year-old blood stains) is from a specific person. It also allows for positive identification of parent-child links. Discovered in the early 1980s, genetic fingerprinting reached the market in 1987 and has already been used by the immigration authorities and the courts of the United Kingdom to settle paternity claims and to solve crime cases (including one case of rape where the new technique led to the positive identification of the culprit).

The long-term human benefits, and the legal and social changes which a radical transformation of crime detection and forensic medicine through this and other methods will bring about are difficult to fathom, but will be far-reaching. Potentially even more far-reaching could be the applications of gene-mapping and gene-sequencing results to inherited and other diseases where genetic predispositions may play a role (e.g. diabetes, schizophrenia, cardiovascular diseases).

The idea of gene therapy is not entirely new, but only with the development of recombinant DNA techniques did the prospect of curing genetic disorders come within reach. The interest in this is all the more justified as the list of disorders recognised as being of genetic origin has been constantly growing, from 400-500 before 1960 to 1 487 in 1966, 2 811 in 1978 and over 3 000 in 1982[2]. For most of these there is no real therapy at the present time.

Some of these disorders, which together affect about *5 per cent* of births are due to mutations by a single gene, others to several genes and others again to anomalies in the number of chromosomes. They may predominate in particular ethnic groups.

Ethical questions in gene therapy have been discussed for more than a decade. Have we the right to introduce changes into the human genome which could be passed on, thereby influencing the very evolution of mankind? Are we ultimately to alter such characteristics as intelligence, physical strength, colour blindness, short-sightedness, etc.?

Other ethical issues arise with the possible prevention of hereditary disorders through prenatal diagnosis, when that is possible. As understanding of human genetics improves, the small number of disorders which presently can be detected in the womb will increase.

In view of such concerns, and also of some scientific and technical difficulties (Chapter I, 3*c*), experiments in gene therapy are presently confined to somatic cells. Germ cells are not experimented with because any change to those could be incorporated in the heredity of the descendants of the patient concerned.

Difficult questions remain to be solved: how will genetic knowledge be used by employers, insurers? Existing medical ethics can answer some of these questions but others will require reflection on the part of governments and the development of a new legal framework.

It is clear, however, that our understanding of the mechanisms of life and the causes of disease will keep growing through the study of genes, revolutionising the underlying concepts in this still very new field. To judge from past experience, future progress in this field might again be distinguished from other technologies by the rapidity of scientific and technical developments which up to now, has often confounded the forecasts of experts and observers.

2. Problems of measuring quality

At the beginning of the 1980s, an OECD report which analysed the impacts of science and technology in a changing economic and social context concluded that macro-economic analysis as well as economic and research policies of the future would have to give much more attention to the quality of life impacts of technology[3]. The development of biotechnology since then has borne out the timeliness of that recommendation which, however, has had little perceptible effect on national policies or on economic and statistical analysis.

Various problems continue to impede a statistical assessment of quality. For some intangible quality of life impacts, quantitative measurement is conceptually impossible, except maybe through "happiness polls" which would measure whether and how these impacts are perceived by the public.

At the other end of the spectrum are the quantitative changes which are traditionally being measured by economic statistics.

Between the two, however, stretches a broad zone of quality impacts where, at least conceptually, numerical measurements are possible. Health standards and life expectancy, the quality of food and nutrition, the quality of the environment (air, water, soil), the crime rate and personal safety, and other items could be measured albeit by different methods and with varying degrees of accuracy. Many of the quality impacts of biotechnology belong to these categories.

Past discussions on *social indicators* have extensively referred to the opportunities and limits of such measurements[4]. Although work on social indicators is continuing in several OECD countries, progress has been rather slow. There are other priorities in statistical research and it will, in any case, take time before social indicator concepts proposed by the OECD and others are tested and accepted in Member countries.

One of the initial problems was the demand for a single "quality of life index" which soon enough turned out to be impossible. For the time being, the search continues for at least some *"synthetic" indicators* which would summarise quality developments related to important sectors or social concerns, such as public health, or poverty. Progress here is held up by the difficulty of statistically capturing the crucial "distributional" aspect of social or quality changes. Improvements in the health, nutritional or housing status of the population are primarily significant by reference to specific age, ethnic, regional, social, professional or other groups.

A perhaps more readily implementable approach to assessing changes in quality of life would be through better service statistics. Social indicators and service sector statistics, of course, do not measure exactly the same, but new technologies embodied in improved services are generally expected to improve quality of life, if only for the users of the services, and such improvements in services can, in principle, be quantified and measured. For example, improvements in the cost-effectiveness, speed and quality of transportation during the last few decades are being measured and can be found in economic statistics. In the same way, the improvements in new health-related diagnostics, particularly by monoclonal antibodies, which are faster, safer, more precise and cheaper than most earlier methods, represent great improvements in a service and thus, are in principle, measurable.

Unfortunately, the relative backwardness of statistics relating to the service sector of OECD countries does not make this an easy proposition for the time being. While the service sectors contribute in many OECD countries for up to 70 per cent of the GDP, their statistical measurement lags, in detail and methodology, widely behind agricultural and industrial statistics.

In the absence of improved social indicators or service statistics, the only numerical measure for the importance of new diagnostics are sales and market forecasts. Sales of monoclonal antibodies in 1987 amounted to several hundred million dollars, and are expected

Table 1

FORECASTS ON SIZE OF WORLDWIDE FOR BIOTECHNOLOGY DERIVED PRODUCTS

(In millions of dollars)

	Year	Total	Pharmaceuticals and health care	Chemicals	Agriculture and food processing	Energy
Business Communications Co.	1982	59	26			
	1990	13 000	12 600	270	430	
Robert S. First Co.	1985		1 400	250		
	2000		43 000	8 200		
Genex Corporation	1990	10 000				
International Resources Development	1985	520				
	1990	3 000				
International Planning Information (UK)	1990	4 500				
	2000	9 000				
Arthur D. Little	1990				2-4 000	
	2000		23 000			
Policy Research Corp.	2000		5-10 000		50-100 000	
Predicasts, Inc.	1985		1 120		6 200	
	1995		18 600		101 000	
T.A. Sheets and Company	1980	25				
	1990	27 000	2 900	5 100		9 400
	2000	64 000	9 100	10 600	21 300	16 400
Strategic, Inc.	1990		5 000		4 500	
	2000				9 500	
U.S. Congress, Office of Technology Assessment/Genex Corporation	1990					
	2000	14 600				

Source: *High Technology Industries, Profits and Outlook: Biotechnology*, US Department of Commerce, International Trade Administration/Genex Corporation, Washington, D.C., 1984.

to reach 1 billion dollars in the early 1990s, if one includes the monoclonal antibodies for therapeutic use which are in development. By then, monoclonal antibodies will have helped countless numbers of people in all countries, improving health and saving lives. However, sales figures are a poor measure of the human, social and indirect economic benefits of what is certainly one of the great inventions of the last quarter of the 20th century.

As the problems of measuring quality-of-life changes are today widely recognised, the increasing importance of quality considerations in economic discussions and policy may, during the coming decade, bring about improvements in service statistics or in social indicators.

For the time being, and in the absence of such improvement, what is true for monoclonal antibodies is true for all biotechnology products. The only way to indicate their importance, actual or potential, is by sales or production figures. However, precise data on current sales and production have until this very day (1988) been extremely scarce and fragmentary[5].

In contrast, global market forecasts for biotechnology products have been numerous, but so have doubts about their reliability and relevance[6].

Table 1 compares results of eleven reports provided by different institutes during the early 1980s. This shows an extremely wide range of forecasts, with an order of magnitude separating the least optimistic forecast of 9 billion dollars for the year 2000, from the most optimistic one of over 100 billion dollars.

Another quantitative assessment of the potential health impacts of biotechnology is to estimate the worldwide populations affected by diseases which could be treated by new biotechnologies. For people presently affected by parasitic diseases only, the figures are in the hundreds of millions (amoebiasis 350 million, trachoma more than 400 million, malaria 100 million, amongst many others)[7].

But again, such quantitative estimates remain unsatisfactory. Not only do they ignore the problems of delivery and financing of treatment, but they also fail to reflect the most essential element in the impact of biotechnology, namely that the forthcoming revolution in the understanding of life and health is likely to affect, in the end, the treatment of all, or nearly all, diseases.

I. MAJOR TRENDS IN SCIENCE, TECHNOLOGY AND APPLICATION: THE NEXT TEN YEARS

1. Developments since the 1982 OECD report, *Biotechnology: International Trends and Perspectives* (A. Bull, G. Holt, M. Lilly)

During the past years there has been a growing realisation that new biotechnology represents probably the third and most dynamic technological revolution of the 20th century preceded only by nuclear energy and information technology. Among the fundamental aspects of this revolution are: *a)* the development of recombinant DNA technology based on the powers of gene cloning and splicing which allow for the production of large quantities of DNA and for the expression of DNA towards the production of rare proteins, *b)* hybridoma technology allowing for the fusion of specific antibody-producing spleen cells with myeloma cells to produce large quantities of pure antibodies, and *c)* instrumentation for the microsequencing of proteins and DNA and for the synthesis of oligonucleotides and peptides.

These and other powerful technologies based on earlier fundamental research will have an increasing impact on the world's major problems of disease, malnutrition, energy availability and environmental deterioration.

The 1982 OECD report has been remarkably accurate, and its concern about the realisation of products for the market were well founded, although the rate of product introduction has increased since 1986 in the health industry. Since 1982, excellent progress has been made in plant gene engineering, and in cell and tissue cultures, but much remains to be done in gene transcription, translation, processing, promoters, tissue and developmental gene regulation and gene induction. Good progress has also been made in targeting gene products in plants. Improvements in plant science funding have occurred in many OECD countries and may have been influenced in part by publications such as the 1982 OECD report.

Exploration of various microbes as alternate hosts to *E. coli* for gene expression has moved quite well in yeast, *Bacillus sp.*, and *Streptomyces sp.* but less so in *Lactobacillus* and in fungi, anaerobes and thermophiles. Interestingly, mammalian cell culture has gained in significance as interest in various glycoproteins has developed. The use of new expression vectors such as *baculovirus* has unexpected advantages for the production of therapeutic molecules, including a first candidate for the production of a possible AIDS vaccine, much beyond what would have been predicted only a few years ago. Other technologies, such as *in vitro* gene amplification, are finding applications beyond those for which they were initially developed.

A definite upswing in biochemical engineering has occurred with increased training of chemical engineers and greater recognition of problems beyond fermentation, including fermentation control systems, online monitoring, analytical control systems, cell processing, product separation and purification and scale-up issues. A new appreciation for the importance of chemists and engineers in biotechnology has developed. Much progress has been made in isolation and purification techniques for biological macro-molecules (peptides and proteins).

University and industry relations have grown in an orderly and successful manner with increasing agreement that both sides can work together in a mutually beneficial partnership.

In general, raw materials have not been a major issue. For some, mainly European countries, raw material prices, particularly sugar prices for fermentation, continue to present problems. In contrast, the decline in oil prices has tended to reduce biotechnology efforts focusing on the production of energy sources and chemicals.

Economic considerations and analyses have replaced some of the early euphoric hopes for products derived from biotechnology. This has resulted in a significant focus on high value-added products rather than commodity products. Combined evaluations of gene engineering, mutagenesis, fermentation optimisation and downstream processing have intensified as economic dictates have become more obvious.

Risk assessment and safety issues have remained a concern, and are currently under study, particularly the introduction into the environment of new varieties of micro-organisms, plants and animals. Initial field introductions have been approved.

There are other complementary and synergistic technologies that will influence the economic trends and impacts of biotechnology. Among these are the more traditional techniques of plant breeding, somatic cell culture and information technologies (computer science, instrumentation, etc.).

New opportunities for servicing the growing activities in biotechnology have led to many new businesses which provide computer support for molecular biology, equipment for nucleic acid and peptide synthesis and analysis, or contract mammalian cell culture production, to mention only a few advanced activities. Not all of these ancillary industries belong to high technology sectors, but they are nevertheless making a positive contribution to innovation, economic development and employment.

2. Current and future perspectives: major sectors of application

a) *Pharmaceuticals (Drugs and human health care)*

The new biotechnology has clearly had its earliest and greatest impact on the pharmaceutical and health care industry. Already products of this industry have emerged in the form of insulin produced by bacteria for use in the treatment of diabetes, several interferons for the treatment of cancer and leukemia, human growth hormone for the treatment of pituitary dwarfs, tissue plasminogen activators, a natural proteolytic enzyme used for the dissolution of blood clots, and a hepatitis B sub-unit vaccine. Other important new products are the more than 200 diagnostic tests uniquely capable of detecting diseases including chimeric (that is humanised) monoclonal antibodies, some of which will also be used in therapy.

The development of natural protein and peptide substances into useful therapeutic drugs, in addition to those mentioned above, includes materials such as atrial peptides, interleukins 1, 2, 3, 4 and 6, tumor necrosis factor and growth factors, such as insulin-like growth factor, nerve growth factor, epidermal growth factor, platelet-derived growth factor and fibroblast growth factor. Studies with these natural products have indicated that their systemic administration may exaggerate undesirable activities normally masked when the substances are produced endogenously. This finding has initiated research on second and third generation products made by taking advantage of genetic engineering and protein engineering technologies. These new generations will range from the application of site-directed mutagenesis to create specific point mutations in a given gene, to much larger-scale modifications created by shuffling different domains of genes in order to generate hybrid molecules and molecules with totally different physical and chemical attributes. The result will produce more specific drug targeting, decreased degradation rates and higher specific activities.

Protein engineering of the type just mentioned coupled with powerful tools, such as X-ray crystallography, 2D NMR and molecular modelling will lead to new structure function insights and ultimately to the design of stable peptidomimetic drugs for oral application.

The general trend will be toward disease diagnosis (immuno-tests, gene-probes, biosensors, etc.) and prevention (vaccines) rather than cure, although this will require the continual generation of new knowledge related to the etiology and pathobiology of disease. There will certainly be a major upswing in the development of new vaccines.

Immunology is one of the areas which will be most affected by new biotechnology. One of the newer areas for applications of immunology that will be further developed is the creation of antibodies to transition state enzyme substrates. It is hoped that these antibodies themselves will then possess enzyme activities that can be further modified by mutagenesis and hence, lead to novel enzyme activities of use both to health care and to industry.

While gene therapy has been viewed as a significant new possibility of dealing with a number of genetically inherited diseases (see Introduction), progress has not been rapid and awaits a further understanding of cellular transformation and gene expression. Large-scale application is not likely to occur in this century.

Acquired Immune Deficiency Syndrome (AIDS) has dramatically influenced society and become a problem of severe proportions. New biotechnology has already facilitated the understanding of this disease by the generation of monoclonal antibodies, DNA probes and genetic structural analysis, thereby enhancing our ability to detect the virus. Simultaneously, numerous approaches to dealing with this disease have been initiated utilising biotechnologically-generated products. Vaccines, viral inhibitors and immune modulators are being actively pursued and hopefully will provide a solution before the end of the century.

Finally, an entirely new challenge has arisen to deal with the delivery of protein/peptide drugs and therapeutics. These large molecule drugs cannot be absorbed orally without being degraded, and injection of large molecules has also proven to be problematic and occasionally followed by side-effects. Delivery problems, including for vaccines, have led to delays compared with the expectations of a few years ago. Therefore, drug delivery technologies, both for men and animals, have become a major area of interest, mobilising already several dozen specialised companies. New and novel approaches for drug delivery will be required and will draw heavily upon the advanced physical, chemical and modern biological tools mentioned.

Modern biotechnology therefore has the potential to change the technology paradigm of health care and the pharmaceutical industry, to possibly improve cost control of health care, and to enormously enhance the quality of human life.

b) *Agriculture and forestry*

Agriculture is one of the largest economic sectors throughout the world and one where the stakes of new biotechnology are very great. In the fields of plant and animal agriculture, biotechnology will improve food production by increasing the growth rates and growth efficiency of animals, and plant geneticists as well as molecular biologists will use transformation technologies to create plants resistant to diseases, insects and herbicides and plants capable of surviving in environmentally harsh climates. Genetically engineered microbial organisms to control plant pests and influence nutrient uptake will also be developed. It may yet be too early to predict when such biological means of disease control and stock improvement will lead to a reduction in the use of chemicals in agriculture.

Engineered microbial pesticides have been developed using the delta endotoxin gene from *Bacillus thuringiensis* and transposable elements to insert the gene into the chromosome of root colonising bacteria such as *Pseudomonas fluorescens*. These engineered bacteria will be able to colonise the roots of developing plants and protect them from attack by insects such as the corn root worm. Other genes will be found to protect plant roots from attack by fungal and bacterial diseases as well as by soil borne nematodes.

Significant advances in plant genetic engineering have occurred since 1985-86 and demonstrations under field conditions were recently completed showing that tomatoes can be engineered to be viral disease-resistant, insect-resistant and herbicide-tolerant. There is little doubt that the momentum in this field is accelerating and that advances in the development of more rapidly growing plants, particularly trees, the creation of salt and stress-tolerant plants and plants whose seed qualities are improved (improved oil properties, starch characteristics, protein quality) will be forthcoming within a decade. One of the key breakthroughs anticipated in the short term is the capability to transform monocotyledonous cereal grains such as maize, wheat and rice.

Other types of technologies are also receiving attention in the plant biotechnology areas, particularly novel diagnostics for plant diseases and plant viruses using nucleic acid hybridisation techniques. Biochemical and molecular fingerprinting of plants is also evolving with the application of isoenzyme diagnostics as well as restriction fragment length polymorphism, a very powerful technique for the relative classification of genetically related materials.

Understanding of major plant biochemical pathways and gene regulation will progress, though perhaps not as rapidly as desired, and provide the basic knowledge needed to isolate useful single and multiple gene traits for insertion into economically important crops. Without doubt, plant breeding will be accelerated by advances in somatic cell culture (plant tissue culture) and genetics as well as by more sophisticated chemical/genetic regulation of male fertility.

The development of animals for agriculture will be influenced by current developments in that field, *per se*, as well as by activities in the health care arena. The latter will facilitate the translation of basic finding from animal models for human myopathies and pathologies to an understanding of health problems in domesticated animals, and occasionally vice-versa.

The development of transgenic animals, while moving rapidly at the experimental level, is already demonstrating the need for a greater understanding of the influence of gene position within a chromosome (context DNA) and the precise directed insertion of genes which will require a greater understanding of homologous recombination.

A key area of agriculture already benefitting from biotechnology-generated products is the livestock industry. Field trials have indicated that bovine somatotropin (BST), a naturally occurring protein in cows, can be supplemented with biotechnology-derived somatotropin to increase milk production and improve feed efficiency leading to more milk for the same amount of feed. A related hormone produced in the pituitary gland of pigs has also been demonstrated to enhance the growth rate and improve the feed efficiency of pigs while concomitantly reducing the amount of back fat. New and improved vaccines are being produced for foot and mouth disease, scours, shipping fever and other diseases of domesticated animals. The understanding and use of retroviruses for the creation of transgenic animals is advancing and may provide a means by which animals will be born resistant to various diseases and eventually endowed with enhanced growth rates and feed efficiencies.

In animal reproduction, sex-specific semen may well enhance the business of embryo transfers by producing a greater concentration of embryos of desired sex, although this has also led to concerns that sex-specific semen could be used in human reproduction. Research focussing on major reproductive hormones, such as luteinising hormones, folicle-stimulating hormones and gonadotropin-releasing hormones will both benefit animals and produce spin-offs for various aspects of human fertility.

A relatively new concept in animal agriculture is the study of repartitioning agents, agents that increase the proportion of muscle tissue relative to fat. This area is likely to receive considerable attention during the rest of this century and should lead to new insights in controlling the quality of animal food products. That such agents could also be used in human nutrition might, however, become a source of concern.

Finally, the study of natural appetite modifiers, some of which are peptides, such as opiates, endorphins and enkephalins, will develop a better understanding of the biological mechanisms involved in appetite modification.

In many of the above-described developments, fundamental molecular biology, with particular emphasis on receptor biology and the regulation of hormone/receptor systems, will be pursued. As anticipated for the pharmaceutical industry, this transition will ultimately lead to the development of organic molecules that replace or augment the first generation of biotechnology-based animal products.

Lastly, an area of rapidly developing interest is the use of animals as bioreactors to produce rare proteins. For example, the tissue plasminogen activator gene has been engineered to create transgenic mice that produce and secrete tPA into the mammary glands. By simply milking the animal, the product can be isolated and purified. Similar studies are being conducted with larger mammals, and this novel bioreactor may find utility for the production of complex proteins that require unique processing, such as glycosylation. The economics of manufacturing rDNA products by this means has yet to be established.

c) *Foods and feeds*

Many aspects of the food and feed industry could benefit from biotechnological advances made in plant and animal agriculture as described earlier. Significant quality changes can be anticipated in food products derived from these sources. Examples of these advancements will involve the molecular biology of wheat breeding for the improvement of bread making qualities. These qualities are influenced by the percentage of protein in the grain and the relative proportions of specific proteins, such as the gliadins which allow viscous flow and impart extensibility to dough and glutenins which impart viscoelasticity. Furthermore, it will be possible to modify the nutritional quality of plant proteins by genetic manipulation designed to increase lysine and methionine content of specific seed storage proteins.

Biotechnologically-derived and improved enzymes for food processing, such as bovine chymosin (rennin) used in milk-clotting, have already undergone first generation developments. Novel bioreactors for the development of soy sauce using a plant cell separation enzyme and a new species of plants derived through cell fusion are receiving exceptional support. The further modification of these enzymes through protein engineering will lead to the potential industrial production of foods under conditions that are more efficient and cost effective. This area holds great opportunities for the food processing industry.

The genetic modification of food using genetically engineered lactic acid bacteria represents yet another area for growth and opportunities. These bacteria find utility in the production of thickening agents, natural food preservatives and enhancers of flavour development. These bacteria also represent a safe production host for the manufacture of a variety of food grade products such as chymosin for the manufacture of cheese. Significant advances can be expected in other food grade microorganisms, such as yeasts (for example for low-calorie beer), bacteria and fungi. The use of these organisms for the production of important food technology enzymes and in various fermentation processes will be beneficial.

Studies of sweeteners and flavour enhancers have involved the cloning of genes for thaumatin, a protein isolated from the fruits of a West African shrub. Some five different clones encoding different forms of this protein exist and further studies should lead to a molecular understanding of sweetness perception utilising these molecules and the tools of protein engineering. Clearly, the food products derived from the applications of fermentation technology, enzymology and food microbiology will benefit from advances in basic molecular biology and biotechnology.

Lastly, the food industry will benefit from the development of probiotic molecules used as feed additives to enhance the digestibility of feedstuffs.

d) *Chemicals – speciality and commodity*

Many important industries are actively using fermentation to produce industrial chemicals. They range from the manufacture of potato starch, gluconates, lactic acid, steroids, antibiotics, amino acids to a wide variety of enzymes. Amino acids will be among those products benefitting most from advancements in genetic engineering, host strain modification and improvements in the bioengineering sciences (fermentation, analytical methods, reactor design, etc.). The major trend in these instances will be the application of fundamental genetic engineering to cost reduction in the production of a variety of amino acids, such as phenylalanine, aspartic acid, tryptophan and lysine. Efforts will also be devoted to the development of processes for the manufacture of D-amino acids by the use of stereo specific enzymes created by genetic engineering.

Biotransformation, which may include fermentation but is somewhat more specific and sophisticated, will find increasing use in the modification of molecules. Biotransformation is the modification of organic substrates using enzymes and other biological systems. It is likely that chemists will use enzymes for synthesis instead of classical organic reagents. Advantages of such approaches would include reduction of side products, thereby minimising pollution problems and maximising yields. It should be noted that such reagents (enzymes) used in organic reactions will help the chemist in the laboratory and may well see commercial application in more sophisticated and high value products.

Protein engineering will provide a major tool for the improvement and development of industrial enzymes, an opportunity which is already actively pursued by industry. Examples will include enzymes with greater stability, unique physical properties (e.g. activity in nonaqueous systems) and novel applications (e.g. immobilised membrane reactors, specific chemical processing, sensor application and diagnostics). New microorganisms such as *Pichia stipitis* for the industrial production of ethanol from hemicellulose hydrolysates and methanol-utilising microbes for production of commercially interesting products will receive particular attention. Besides yeasts and bacteria, the industrial use of fungi, notably *Aspergillus species*, are also receiving increasing attention as potential hosts for the manufacturing of products. Much fundamental work will be required to develop appropriate transformation systems and vector expression systems for the chemical industry. This industry will need a strong scientific foundation to support its use of biotechnological methods.

Modest progress may be expected in the development of speciality polymers, although at the present time this area is receiving little attention. Efforts in the development of cloning systems for the expression of high quality polymers, such as silk, and biodegradable polyesters, such as polyhydroxybutyrates, appear to be receiving increased attention. The basic factor governing the success of these future endeavours will be economic, in view of their unique niche-type properties.

Regarding feedstocks for biotechnology applications, it is likely that sucrose, starch, methanol, paraffins and lignocellulose will be potential replacements for the commodity chemicals currently produced from petroleum. However, with present oil prices, one should not expect a significant impact of biotechnology on commodity chemicals in the next decade.

e) *Environment and mining*

While much has been written about the environmental uses of biotechnology, including about biofilters which play an increasingly beneficial role in waste-treatment plants, many major commercial developments for pollution control, waste treatment, oil recovery and mineral leaching or enrichment will be slow in the next decade. Nevertheless, significant progress can be expected in the understanding of the genetics and physiology of microbes engineered to deal with these problems.

Important future applications for biotechnology in the environment will include the areas of waste water purification, groundwater contamination and the recycling of important chemical materials such as sulphur. Progress in this area will be economically influenced by the coupling of reactor design and chemical engineering principles with the genetic modification of appropriate microbes. Bacteria will be engineered to enhance the metabolism of sequester-specific toxic wastes. While naturally occurring bacteria have abilities to degrade a variety of specific chemical agents, modern genetic engineering and selection tools will be able to enhance the abilities of these microbes to be more efficient and cost effective.

Microbial desulphurisation of coal and the leaching of heavy metals from solid waste treatment of various industrial process streams are also receiving attention but will probably develop at a more sluggish pace. The reason for this slower realisation of possibilities is the economics of raw materials, processing costs, capital costs and technological constraints which will work against biotechnology solutions until the cost/technology parameters shift.

3. Current and future perspectives: some key areas of research and generic technology

a) *Biosensors*

This area covers the application of biological functions to the field of electronics using molecular recombinant DNA technology and physical chemistry. Biosensors may be defined as analytical devices which consist of biological molecules and electronic devices. Some examples are:

Enzyme Sensors – used for food analysis and food processing controls;

Microbial Sensors – using living microbes for process control and environmental analysis;

Immunochemical Sensors – which are under investigation. No products exist at present. The technology will allow the detection of 10^{-12} grams per ml. using labeled antibody systems.

Most biosensors developed to date are based on electro-chemical devices. A new wave of technology involving silicon fabrication technology is emerging for the production of micro-biosensors or transducers. For example, the field effect transistor can be used as a transducer for biosensors. Thus, changes in pH in solution when monitored on immobilised gate areas on a transistor can be used to detect the substrate of an enzyme due to pH changes.

Microfabrication technology of silicon can be applied to the fabrication of hydrogenperoxide and oxygen electrodes that are 100 micrometers wide and four millimeters long.

Lithographic methods can now be applied to immobilize enzymes onto microelectrodes and transistors. Furthermore, one can integrate the microelectrodes or transistors into a single silicon chip using ordinary established technology to make a multifunctional biosensor.

Thus, simple, cheap and disposable microbiosensors can be produced by mass-production technology and one can envision the development of ordinary biodisposable sensors for health care applications at home. Similarily, implantable biosensors will also be feasible using integrated circuit technology and biocompatible polymer membranes. These would find use in monitoring a variety of organs within the body, such as the pancreas or the kidney.

b) *Bioelectronics – biochips*

Molecular electronics and biochip technology have been discussed in the context of electronic or computer components that could mimic living cell capabilities to store and retrieve information in a dense form. In the scientific community, there are proponents as well as sceptics regarding the possibility of building a computer made up of proteins and other molecules functioning as electronic devices. Nevertheless, in recent years the momentum of activities in the area of bioelectronics has intensified despite the technological obstacles. While the dream of building an organic computer may not be feasible even during the 21st century, it has been predicted that efforts in the area of molecular electronics should lead to the development of new materials, techniques and products. It is apparent that Japan and Germany, along with several US laboratories, are focussing resources on the development of biochips.

Protein engineering research directed toward the development of biosensors may provide a synergistic technology toward the evolution of molecular computational systems. Conceptually, the approach will involve the use of organic molecules as wires and switches imitating the function of silicon components. Organometallic compounds that conduct electricity (porphyrins and pthalocyanines) are foreseen as taking the function of wires. Conductive polymers such as polyethylene and polyphenylene sulfides are also under study. In the case of switches, efforts are focussed on evaluating optical switching devices based on bacterial rhodopsin and tetracyanoquinodimethane.

Soliton switching represents yet another concept for molecular electronics. This mechanism involves the use of electrically conducting polymers, the conductivity of which is due to wave-like perturbations that propagate down a chain of polymer without dissipating energy. As the soliton travels it changes the electronic configuration of the polymer (single and double bonds exchange places) and this conformational change results in a flow of electrons through the molecule. Lastly, Langmuir-Blodgett film technology is being developed to create biochip substrates. It is likely that fundamental research in immunochemistry, neurochemistry and receptor biology will also provide useful information for the development of biocomputers.

Biochip research has just begun and perhaps the attention devoted to it has been somewhat exaggerated regarding current possibilities. Our lack of knowledge and the attendant technical obstacles suggest that biochips must be considered a dream for the moment and must await totally new synthetic approaches to make real progress.

c) *Human genome research*

Japan has initiated a major programme to map and sequence the human genome, and other countries are planning similar efforts. Extensive evaluations for launching such a programme in the United States are under study. The programmes are aimed at providing a data base essential to facilitate research in human molecular genetics and should significantly affect approaches to diagnosis, control and prevention of disease.

The major thrust of these investigations over the next decade will require high resolution genetic linkage maps of the human genome, collections of ordered DNA clones, the development of a series of complementary physical maps and ultimately complete nucleotide sequences. In addition, a number of comparative genetic approaches will be required in parallel to augment interpretation of the human genome map.

This mammoth task will provide entirely new opportunities for the creation of scientific and technological breakthroughs required for separations of large pieces of DNA, cloning of large pieces of DNA, separation of intact chromosomes, cloning complementary copies of expressed genes, cloning large DNA fragments, purification of large DNA fragments and the automation of these various steps. There will also be a concomitant need for at least an order of magnitude improvement in the scale and speed of DNA sequencing and for improvements in sequencing large nucleotide sequences.

Finally, the management of information and materials for such a massive programme (the human genome contains three billion base pairs) will require a mechanism for the collection, storage and analysis of data as well as for making this data accessible to researchers. The automation of such a programme has been initiated in Japan and complementary activities have been initiated in the United States and the United Kingdom.

d) *Biological function research*

Advances in the expression of genetic information, molecular recognition and response, cellular transduction, morphogenesis and energy conversion from the molecular/biochemical viewpoint will undoubtedly contribute to a growing body of fundamental science utilising the tools of biotechnology and molecular biology.

In addition, advances in immunology and immune modulation will provide a molecular scaffolding for the understanding of molecular immunology and fundamental higher nervous system functions, such as perception and recognition, memory and learning and motor control. The line between immunology and neuroscience will probably be blurred as these fields converge in a synergistic manner. The application of these advances to the understanding of auto-immune diseases and degenerative neural diseases will enhance the prospect for solutions to diseases common to an increasingly older population.

e) *Other research and technology areas*

While little has been said about biochemical engineering, bioreactor design and large-scale mammalian cell culture, these areas will receive significant and growing attention in the next decade due to their pivotal role in the applications of biotechnology to the various fields described earlier. All of these areas will require additional new breakthroughs and fundamental understanding if they are to be economically efficient in the scale-up for commercial applications.

Enormous new opportunities will continue to surface in providing services to support the basic advances and their subsequent applications to societal needs. Thus, the number of companies supplying instrumentation and services to both academic and industrial complexes will surely grow. The development of providers of reagent supplies, cells, diagnostics, etc., along with consortia that provide services such as X-ray crystallography, nuclear magnetic resonance spectroscopy, high speed computational analysis and molecular modelling capabilities can also be anticipated.

II. CURRENT POLICIES OF INDUSTRIAL FIRMS IN BIOTECHNOLOGY

The present Chapter will examine current plans and opinions of industrial firms, including less advanced firms, with regard to biotechnology. It summarises an unpublished report by Prognos AG, Basle to the OECD[72] and also takes into account supplementary information provided by other experts. The gap between the very promising scientific opportunities and trends (Chapter I), and the numbers of products currently on the market may point to some of the difficulties encountered by biotechnology. These include skilled manpower problems, regulatory delays and also insufficient market demand which, in certain sectors (food), has played a significant role.

1. Methodology

Industry plays the pivotal role in transforming new biotechnology into an economic force, and it is assumed that present industrial patterns and policies are already an indication of future economic developments. To gain insight into these policies, 94 industrial companies in 17 OECD countries were interviewed in 1986 and 1987 for the purpose of this study (List in Annex 2).

The question of definition, extensively discussed in the Preface to this report, has been a major problem and has taken considerable time during the interviews. Company definitions vary, often depend upon the interviewed manager, and are still evolving. The present Chapter speaks of "classical biotechnology" (CBT) which is thousands years old, "modern biotechnology" which evolved during this century through a process of continuous technical improvement, and "new biotechnology" (NBT) based on the discoveries and tools of the last ten years, including genetic engineering.

Although a structured questionnaire was submitted beforehand, the interviews did not rigorously adhere to it, and in general did not insist on replies of the "yes-no" type. Most companies did not reply to all questions. Furthermore, replies on complex matters often required interpretation, occasionally leaving some uncertainty about the correct categorisation of some companies. All this explains why the percentage figures calculated for this chapter are often approximate and may sometimes appear scattered. In most companies, the managers interviewed were responsible for R&D, and in a number of cases it was possible to interview more than one person.

Although the 94 companies interviewed were not chosen on the basis of a statistical sampling technique, their distribution is sufficiently broad and their share of production sufficiently important to allow for an assessment of trends in the biotechnology industry.

In addition, discussions took place in many countries with industrial associations and government agencies and ministries.

The *geographical* distribution of the companies interviewed is:

21	USA/Canada	43	EEC
15	Japan	15	Europe non-EEC

It is more difficult to give distribution by *sector of activity* because many companies are active in several sectors. Preference was given to companies from sectors where traditional and modern biotechnologies already play a role. Thus, the large majority of selected companies belong to the *pharmaceuticals*, *food and feed*, *agricultural* and *chemical* sectors, with a smaller number active in the *environment* field and in the manufacturing of *biotechnology equipment*.

A distribution by *size* shows:

31 large multinationals, some with considerable R&D budgets and mostly active in several if not all of the above-mentioned sectors;

48 small and medium-sized companies, focusing more on national markets, and some with only limited R&D budgets;

15 small biotechnology "start-up" companies, spending up to 100 per cent of their budgets on R&D;

52 of the 94 companies have already had experience in classical or traditional biotechnologies.

2. Reasons and strategies for moving into biotechnology

a) *Main results and sector differences*

Of the companies interviewed:

nearly 100 per cent indicate interest in biotechnology (BT);
approx. 90 per cent have plans in biotechnology (which means that 10 per cent have none, though they may profess interest in principle);
approx. 65 per cent have plans specifically in new biotechnology (NBT);
approx. 50 per cent have development projects or activities in biotechnology;
approx. 30 per cent have initial research projects in biotechnology.

Important variations can be found between sectors of activity. Table 2 compares the number of companies with no biotechnology (BT), or with classical biotechnology (CBT) experience to the number of companies which have experience, plans or projects in new biotechnology (NBT):

Table 2

NUMBER OF COMPANIES WITH BIOTECHNOLOGY EXPERIENCE OR PLANS, BY SECTOR[1]

Sector:	I: No BT experience	II: CBT experience	III: NBT experience	IV: NBT plans or projects
Pharmaceuticals	23	17	8	29
Food/Feed	17	14	5	4
Agro-Industry	9	6	–	14
Chemicals	18	15	–	13
Environment	2	–	1	8
Equipment	–	–	9	2

1. This is a simplified version of Table 2.2 in the Prognos report (note 72). The numbers add up to more than the total of 94 sampled companies, because many companies with several sectors of activity are counted several times, and within individual sectors, some companies with presently no biotechnology (BT) experience do have new biotechnology (NBT) plans and are, hence counted twice (under I and IV).

In the pharmaceutical and agro-industrial sectors (the latter includes companies manufacturing agricultural and horticultural products, and companies active in animal husbandry, water management and forestry), a large number of companies are already active, or plan to be active in new biotechnology, compared to those which have no, or only classical biotechnology experience. In the food and feed sector, the number of companies which are, or will be active in new biotechnology is very small – smaller than in any other sector – compared to the numerous companies which have no, or only classical biotechnology experience. Apart from insufficient market demand for new biotechnology products, this is also a reflection of the not very innovative nature of the food industry. In the chemicals sector, the number of companies which plan to move into new biotechnology is slightly smaller than the number of companies with no, or with classical biotechnology experience. Nearly all companies in the environment field are, or will be active in new biotechnology, and obviously all biotechnology-equipment manufacturers are or will be active in new biotechnology.

The table also indicates that the number of companies which will develop and use new biotechnology is likely to increase steeply in the coming years, compared to the number of those already applying new biotechnology. This is true for all sectors except food and feed.

Every company has a survival strategy which is primarily based on the attainment of a certain market share and a certain profit. Biotechnology is but one tool used in this strategy. All companies interviewed underlined the need to maintain their position in the market irrespective of their attitude towards biotechnology. All are discussing intensely whether they should put emphasis on marketing strategies or on new product developments. Only five companies are relying exclusively on marketing: all are multinationals with relatively few products and large market-shares in the food sector.

The overriding motive for a company to develop biotechnology is a perceived need to increase know-how in order to improve its market position. Of those which have made a decision in favour of biotechnology (particularly new biotechnology), about 65 per cent indicate that they want to develop *new products*. Screening of possible products is the next and crucial step upon which further activities will depend.

About 25 per cent point to the opportunities for *process rationalisation* as the reason for their interest in biotechnology. Process rationalisation includes improvements in raw material supplies, in waste disposal, purification and numerous manufacturing techniques.

The precise reasons and strategies for developing biotechnology activities depend on the following factors which vary from company to company:

i) Current product lines and fields of activity;
ii) Familiarity with biotechnology concepts;
iii) Innovation oriented top management;
iv) Available R&D potential.

Companies with biotechnology-related product lines (e.g. antibiotic manufactures in the pharmaceutical sector, beer or amino-acid manufacturers in the food sector) and those already familiar with biotechnology concepts show particularly strong interest in new biotechnology. All or nearly all 52 companies which already have some classical biotechnology experience consider it as essential to extend their know-how in the direction of new biotechnology. However, recognition that this is essential is not always matched by real activities or budgets – only three-quarters of the companies with classical biotechnology experience (approximately 40 out of 52 companies) have also made an R&D budget available for new biotechnology. The proportion is particularly high in Japan (all companies but one), but much lower in Europe.

For companies which in contrast to the preceding group, are not already familiar with biotechnology concepts, the presence of an innovation-oriented top management is crucial in the decision to develop new biotechnology. Management creativity does not so much depend on training in biotechnological disciplines but on the capability of decision-makers to see biological and economic factors in combination, and on the availability of sufficient information to these decision-makers.

The implications of company decisions to develop new biotechnology activities vary widely, according to general company outlook. Three types of strategies can be distinguished:

i) Some companies plan to use new biotechnology for modest changes or extensions of their current product lines or for diversifying into product groups where other companies are already competing;

i) Other companies have, thanks to their structure or management competence, built up a capability to link together biological phenomena, technical possibilities and socio-economic issues and needs. This capability has, in some cases, led to novel product lines;

iii) A third strategy aims at large future markets, through long-term accumulation of know-how.

Common to nearly all companies which want to use new biotechnology, is the accumulation of basic and applied know-how together with a continuous examination of market opportunities, efforts to reduce costs in view of the dearth of profitable products, and an increasing tendency towards secrecy.

Thus, companies face two major questions: how to identify new products and how to accumulate biotechnology know-how. Whereas they were rarely willing to answer questions about products and their identification, companies more amply explained their know-how and R&D strategies.

Summing up the results of these interviews, the movement of industrial companies into biotechnology seems to be broadly based and fast, and may be irreversible. Considering that a number of the 94 companies interviewed are not "high-tech" companies and that new biotechnology is still in its infancy, the proportion of those companies with interest, plans or projects in biotechnology, and particularly new biotechnology, is remarkably large. Already 30 to 40 per cent of the interviewed companies see in biotechnology a new generic technology comparable to the computer, requiring a permanent pool of in-house know-how.

Preliminary estimates of industrial R&D in biotechnology in OECD countries confirm these conclusions. For instance, in the US, the NSF estimated R&D expenditures in biotechnology to be approximately $1.4 billion in 1987, with an average growth rate of 17 per cent per year at current prices, between 1984 and 1987. An OTA estimate speaks of $1.5-2 billion in 1987. On the basis of these and other data, total industrial R&D expenditures in biotechnology in the OECD area have probably been between $3 and 4 billion in 1987[8].

b) *R&D strategies*

The increases of biotechnology R&D in industry did not visibly affect the traditional proportions of turnover spent on total R&D. These proportions still lie between 10 and 15 per cent in pharmaceutical companies; between 1 and 8 per cent in food companies; between 2 and 5 per cent in agro-industrial and forestry companies. The new biotechnology start-up companies which often spend up to 100 per cent of their budgets on R&D, are a special category.

However, considerable shifts have taken place within the R&D budgets of companies though it is difficult to gather precise data. However, it is known that, in many drug companies 25 to 40 per cent of total R&D is presently directed into classical biotechnology and new biotechnology projects. The proportions are often smaller in food and agro-industrial companies, but can go up to 100 per cent in companies which have decided to put all emphasis on new biotechnology.

On average, companies plan to increase their R&D spending on biotechnology, not only in absolute terms, but also relative to other R&D expenditures, until approximately 1995. After that, the proportions may decrease again, although opinions about this are often found to vary widely between the interviewed research and marketing managers. Return on investment will decide for how long, and by how much biotechnology R&D will be allowed to increase, particularly in small and medium-sized companies.

A large majority of companies are screening their inhouse R&D potential before they launch new biotechnology projects. However, the indentification of the available biotechnology potential appears to be strongly influenced by varying definitions, and by the personal position, perspective and professional background of the biotechnology manager who, by training, may be a medical doctor, a chemist, a general biologist, a molecular biologist, or an engineer. Varying personal backgrounds can create particular problems because of the relative novelty and interdisciplinarity of biotechnology.

Thus, the introduction of biotechnology can depend upon the general "culture" of a company. Chemical companies, for example, are sometimes still managed by chemists who want to maintain what they have learned in the past. It may take time before they are replaced by a new generation of scientists. In about 20-30 per cent of the inteviewed companies, the formation of a biotechnology R&D "nucleus" is still under way.

Company strategies for developing biotechnology activities are reflected in the strategies for accumulating know-how and in the organisation of R&D. Companies with old, classical biotechnology experience often prefer to retrain their own scientists, and only occasionally acquire new manpower. This reduces both the need to buy outside resources and the danger that know-how might flow to other companies.

For certain large corporations, the single most important component of their know-how and R&D policy is a well planned strategy of co-operation with highly specialised biotechnology start-up companies. This co-operation functions in spite of numerous critical comments of the large corporations against the tendency of their small partners to ask for R&D pre-financing.

Some companies invest in other companies, both for R&D and marketing reasons, if they do not want their involvement in biotechnology to be openly known.

Large and small companies are facing various advantages and disadvantages when carrying out biotechnology R&D. The perception of advantages and disadvantages varies between countries, particularly between the United States and Europe. The reasons are partly semantic; companies of more than 1 000 or 2 000 employees, still called "small" in the United States, are often considered to be large in Europe. Also, in American parlance, "small" companies in biotechnology mean start-up or new technical venture companies, founded by scientists. Their case is quite different from that of the numerous, traditional small and medium-sized companies which in nearly all countries provide the bulk of industrial employment and which Europeans have in mind when they speak of "small" companies.

In the United States, the ability of small biotechnology companies to change their R&D strategies rapidly, has been cited as their main advantage over large companies. In Europe, however, many, particularly traditional small and medium-sized companies face big problems in their R&D strategies. The following problems were mentioned as critical:

- Biotechnology requires a different kind of R&D management to solve new and complex issues. Small companies have much less experience with this than big companies;
- Biotechnology requires interdisciplinarity. While this is relatively new for both small and big companies, the latter have found it easier to correct mistakes and spread losses due to insufficient interdisciplinarity;
- Industrial biotechnology makes high demands on the co-ordination between inhouse R&D and outside sources of R&D, particularly universities. This includes the need for companies to have up-to-date and comprehensive information on all outside R&D sources. Large companies can satisfy this need much better than small ones. Also, they have less difficulty in translating university research results into their own goals than small companies;
- Small companies are particularly dependent upon highly qualified scientists-enterpreneurs with a good market sense;
- Small companies have more difficulties than big ones to find the financial resources (this is true for the United States as well) and the personnel to build up biotechnology R&D activities.

The early years of industrial R&D in biotechnology (until approximately 1985) could be characterised as a powerful, "science push" period. Scientific enthusiasm and competition have led many companies to spend considerable R&D money on the same few products (e.g. interferons), without sufficient regard for markets. Also, insufficient attention was paid to costs, for example of clinical tests and marketing. In the more recent past, companies have become more selective in their research strategies, have begun to take both technical limits and real market needs better into account, and are using more network planning and ex-ante assessments for new products.

c) *Investment and financing*

Industrial investments in biotechnology could only be assessed if biotechnology were clearly defined and separated from other activities. This is the case in some companies which want to single out their biotechnology activities, for capital market, image or political reasons. In many other companies, however, a separation is presently very difficult.

Most investments are still in R&D, as production has barely begun. In general, manpower expenditures amount to more than 60 per cent of investments, hardware expenditures to more than 30 per cent and expenditures for training to approximately 5 per cent. Companies anticipate, for the medium term, a reversal of these relations, that is a trend towards higher capital intensity in investments.

However, there is still a considerable degree of uncertainty about this because R&D results, process rationalisations and the identification of attractive new products will affect the balance between manpower and hardware investments, as well as the balance between self-financed investments and reliance on co-operative ventures or government support. New fundamental research solutions to current technical bottlenecks could have major impacts on production processes and the production and hence investment programme of the biotechnology equipment industry.

Forecasts of industrial sales of biotechnology equipment for research and production show a rapid rise in projected hardware investments worldwide, from approximately $400 million in 1985, to approximately $900 million in 1990, $2 billion in 1995 and almost $4 billion in the year 2000 in 1985 Dollars[72].

Most companies, particularly the larger ones, are financing their investments mainly with company-owned funds. In contrast, start-up companies have often financial problems linked to national peculiarities in capital markets and to conditions set by governments and banks. A continuing critical problem in many countries according to interviewed firms, is the absence of venture capital and of an over-the-counter stock market.

3. Current products and markets

a) *The pharmaceutical sector*

New biotechnology was born in university-based research, embraced by the resulting new gene-cloning companies, and slowly diffused into established pharmaceutical companies. In 1987, more than 100 new products were under development, but only few have already reached the mass-production stage and the market.

Many of the products still under development will have important market niches or mass markets. They will bring noticeable restructuring of health care markets, and make many present products obsolete.

In some sectors, the possibility of mass production based on new biotechnology processes will make oligopolistic market positions possible, with considerable price consequences. Agreements on market-sharing by geographical region, could become necessary as a result of increasing patent fights, and pricing agreements as a result of large R&D expenditures.

Growing demand for information on health relevant factors will widen the field of application of the new diagnostic technologies. No longer limited to the human body, these tests may extend to food, clothing, habitation, work and leisure places and the environment, which would lead to an expansion of the circle of users and to new mass markets (see Chapter V, 3*a*).

b) *The food sector*

New biotechnology developments in the food sector are often spin-offs from the pharmaceutical sector. However, there are, contrary to the drug sector, only few products on the market and in the development phase. Sweeteners based on amino-acids are a prominent example of a product with a mass-market potential.

All food companies interested in new biotechnology face a marketing dilemma. Saturation of the food markets and growing quality demands by consumers require increasing marketing efforts for new products. It is, however, very difficult to make publicity for new biotechnology-based food, and impossible to approach potential food consumers with biotechnological terms (except perhaps Japanese consumers). Thus, food companies hesitate in offering new biotechnology products.

In addition, food companies have not yet clearly assessed what contributions new biotechnology could make in their sector. The biggest opportunities are presently expected to arise from improved processes (enzymes). Another prospect is quality improvement (Chapter V, 4*b*) through emphasis on attractive taste, aroma and freshness - qualities which the traditional food industry has often neglected although it has applied high quality standards to manufacturing processes and safety. In the future, new biotechnology could open up larger markets by focussing on these quality aspects. Marketing studies in whealthy OECD countries have indicated that consumers are ready to pay up to 30 per cent more for quality increases of this type, which could increase food sales in OECD countries by the year 2000, by up to $30 billion.

c) *The agro-industrial and the environment sectors*

Although there are, for the time being, few products on the market, the applications of new biotechnology are more controversial in this sector than in any other.

The potential of new biotechnology to increase productivity growth in agriculture, entails the danger of increasing food surpluses. Some industrialists mentioned, in this context, bovine growth hormone (BGH) as the most important example of a new biotechnology development mistake. Although at least five companies continue to work on BGH, others have stopped development. The possible consequences of a product that could increase milk and meat production on the one hand, and reduce agricultural employment on the other, has made a number of potential manufacturers uncertain, and might lead to more comprehensive technology assessment methods in industry, whereby R&D and marketing costs will be confronted with market perception and acceptance.

The prevention of environmental damage due to traditional fertilizers and pesticides is considered to be an important field of application for new biotechnology R&D.

Some interviewed firms considered that interesting applications for new biotechnology exist in the plant and animal health areas. The numerous overlaps between human and veterinary medicine entail many advantages for the development of animal diagnostics and vaccines. Diagnostics for plants are also considered to be a priority application for new biotechnology.

New biotechnology provides new tools (plant tissue cultures and others) which allow for substantial reductions of the time it takes to grow plants. In the future, public recognition of the increasing recreational, ecological and economic importance of forests, could turn this sector into a primary field of application for new biotechnologies, leading to the development of faster growing, disease resistant and otherwise improved trees.

The complexity of the technical, economic and political interactions in the agricultural applications of new biotechnology, has led to some uncertainty in industry and to a variety of company strategies. Diversification from one crop to another, or from one use of a crop to another, is among these strategies.

In the environment sector a number of companies already produce tailormade micro-organisms and enzymes to decompose soil-polluting agents. The market potential for such products could be considerable. For the time being, safety issues related to the introduction of micro-organisms into the environment, insufficient ecological awareness at political levels, and questions of financing are delaying the growth of this sector (see also Chapter I, 2*e*).

4. General economic questions raised by company replies

a) *Employment*

The employment effects of biotechnology are in each company a direct function of the expected new products and planned cost reductions. Calculated on the basis of presently recognisable biotechnology markets and products, total manpower in the bio-industries will, for the next ten years, barely change.

In fact, the general position of the interviewed companies is that they intend to maintain their personnel at constant levels, or to reduce it through process rationalisation.

However, nearly all companies project increased needs for highly skilled scientific personnel. The question of how to organise and finance these increases is presently one of the main issues in industry. Only 40-60 per cent of the additional scientific manpower demands will be satisfied by increased inhouse employment; the rest will be mobilised through company links with universities and new start-up firms.

This distribution between internal and external resources, with high emphasis on the latter, gives companies considerable scope for trial and error and allows them to reduce overhead costs.

In the view of industry, the American system of "hiring and firing" adds to the flexibility of US companies, compared to European and Japanese companies. However, it is the aim of many companies to reach an optimal degree of independence in know-how, and to build an efficient biotechnology nucleus for new R&D projects and for retraining purposes. This sets limits to their reliance on external R&D sources.

b) *Industrial structure, concentration and the role of small companies*

As indicated under "R&D strategies" (2*b*), some important conditions for success in biotechnology are favouring big rather than small companies. The high costs of research, of marketing, of safety and of patenting, all point in the same direction. On the other hand, the ability of small companies to move rapidly, may put them in a better position than large companies.

There are big differences according to industrial sector and product. In pharmaceuticals, the prevailing conditions strongly favour the big companies. In other sectors (biosensors, monoclonal antibodies, tissue culture), small companies have been, and continue to be successful. In any event, in order to have optimal access to new biotechnology know-how and to reduce overheads, co-operation between companies as well as skillful co-operation management has turned out to be essential. Companies are unanimous in stating that the number of co-operations has grown continuously, and keeps growing. These co-operative ventures have led to a successful, mutually beneficial symbiosis between big and small biotechnology companies.

The amount of co-operation depends largely upon the size of the company which looks for more know-how. Many multinationals are developing a number of products, each with several co-operation partners. Often, the relevant know-how for one product development is not concentrated in one place, which forces companies to combine several co-operations and partners. Large companies tend to co-operate internationally; one of the reasons, is that university know-how often has strong national ties.

Co-operation occurs at the R&D and distribution levels. Many companies have interesting products with a small market niche, but are unable to add up the niches of various countries in order to have a global market of sufficient size and to reach the break-even point. There is a strong globalisation trend in marketing, and the worldwide distribution networks of the multinationals are a critical condition for success in biotechnology.

Nevertheless, due to smaller costs and greater flexibility, the small and medium-sized companies represent a potential of considerable creativity which the large companies are watching very carefully. This applies particularly to the small biotechnology start-ups which offer very specialised know-how, often focus on only one product, and depend upon the R&D strategies of the large companies.

Their average lifetime is relatively short, they maintain multiple, simultaneous engagements with several large companies, and they change contracts frequently. Dependent upon a relatively small number of big companies, they form together with the latter, a dynamic satellite system whereby some of them continue to have a determinating impact on the rate and direction of technical progress in new biotechnology. "Decentralised concentration" seems an appropriate term to describe the industrial structure which is emerging as a result of new biotechnology developments. This structure, as well as the limited markets and the high R&D expenditures encourage companies, big and small ones, to look for arrangements which make sure that possible returns on investment are not jeopardised, including in some cases, an international division of markets.

"Decentralised concentration" based on large corporations, has helped the transformation of biotechnology know-how into products, and has familiarised companies with new patterns of research-oriented entrepreneurship. Decentralised concentration will remain an important form of industrial organisation for new biotechnology developments, because it is the most appropriate solution for large research problems.

However, in the long term it does not have to remain the only form of industrial organisation on which biotechnology is based. In order to exploit the markets which the ubiquity of biological phenomena could create, the entrepreneurial potential of numerous, more traditional small and medium sized companies could be better mobilised.

The capability of small and medium sized companies to react quickly in regional markets and market niches, is an asset in the diffusion of biotechnology. Large companies may ignore such markets or leave them aside for cost reasons.

An improvement of the biotechnological qualification of traditional small and medium sized companies is a precondition of faster diffusion. Moreover, these companies will have to acquire more competence in large-scale marketing because many biotechnology products become profitable only if the national market segments of various countries are added together to form a larger market. New instruments of economic policy and organisation may be able to assist the small and medium-sized companies in this respect.

c) *International flow of goods and the Third World issue*

New biotechnology is a technology of highly industrialised countries, both with regard to R&D requirements and market potentials.

The attitudes and strategies of the interviewed companies relative to biotechnology and Third World issues can be separated into four categories:

i) Indifference seems to be a prevailing attitude. However, Japanese companies have indicated particular interest in the Third World.

ii) Strong interest in the food needs of developing countries can be found in some cases. Several companies hope that microbiologically produced proteins, or more specifically, the relevant manufacturing technologies will find better markets in the Third World than in OECD countries.

iii) Third World needs for new pharmaceuticals, including diagnostics and vaccines, are recognised by everyone, but no one seems to be sure how such developments, deemed to be desirable, could be financed. Some companies suggest the creation of new international financing instruments for this purpose.

iv) Some companies follow the ongoing advances in the genetic modification of plants, and intend to exploit the increasing opportunities to replace Third World crops. In sectors of production where there is a need for raw materials linked to a problem of quality or transportation costs, biotechnology could, in principle, offer to replace old, imported raw materials by new ones produced in industrialised countries, perhaps on marginal or polluted land. Company replies indicate that the pressures of competition do not allow the biotechnology industry to ignore these opportunities. (Chapter VI, 5).

The biotechnology contradiction of Third World countries is that they have the needs and potential markets, but not the purchasing power to make many new biotechnology products profitable for the private sector. The developments indicated under point *(iv)* risk to reduce that purchasing power even further.

There is a danger that this may lead to increasing autarchy. However, the internationalisation of biotechnological markets and know-how, and the global activities of the leading companies, will also exert strong pressures on governments and on small and medium sized companies, to think increasingly internationally. This could stimulate more international scientific and technological co-operation.

d) *Industry versus government expectations*

Industry-government relations in the context of biotechnology are complex, depending upon product, size of company etc. Approximately 75 per cent of the interviewed companies are unimpressed by, and critical with regard to government policies and support. However, the Japanese, contrary to the European and North American companies, did not express criticism.

Complaints from interviewed firms about government policies underline the following problems:

- Too many duplications of R&D expenditures by public authorities, and corresponding lack of policy co-ordination by governments;
- Insufficient fundamental research support, and too slow adaptation of this research to economic needs;
- Insufficient familiarity of public research funds with the daily problems of application (complaint mainly of small companies);
- Insufficient general support (complaint mainly of companies with older products);

Table 3

MAIN ISSUES RAISED BY BIOTECHNOLOGY

Sector	Main issues as perceived by governments and industry	Significance to governments and industry
Health	*Governments* see a major public issue, both in industrialised and developing countries;	High
	Industry sees major markets;	High
Nutrition	*Governments* see food shortages in the Third World, and nutritional deficiencies in industrialised countries;	Low
	Industry sees markets;	Medium
Environment	*Governments* see global resource and pollution problems;	High
	Industry sees an image problem and some markets	Low
Employment	*Governments* see a major social, economic and financial problem;	High
	Industry sees manpower needs of decreasing importance	Low[1]
Profits	*Governments* see a contribution to growth, industrial competitiveness and enterpreneurial motivation;	Medium
	Industry sees its survival and future	High

1. After completion of the interviews for the Prognos report in 1987, shortages of skilled manpower have again become apparent in biotechnology industries. Thus, industry replies in 1988 would probably give a "high" or at least "medium" significance to this issue.
Source: See Notes and References, 72.

- Insufficient support of risk-assessment research;
- Insufficient international co-operation.

In various countries, grave concerns have also been expressed about the regulatory situation. The thrust of these complaints is quite heterogeneous. Also, the meaning of a critique can be different according to whether it comes from a big or small company. Big and small companies have different R&D requirements and different definitions of "fundamental" research.

One of the problems in the industry-government context is that both sides have different perspectives and expectations with regard to biotechnology.

If one looks at five areas of social and economic concern where biotechnology is expected to make major impacts, it appears that government and industry attitudes vary. The meaning of each area of concern is different to governments and industry, the priorities attached different, and so are the expectations regarding the future importance of each area of concern. Table 3 shows these differences.

Some discrepancies are noteworthy. Public health is the one major concern to which both governments and industry attach top priority in the context of their biotechnology strategies.

III. THE DIFFUSION OF BIOTECHNOLOGY THROUGH THE ECONOMY: THE TIME SCALE

1. Introduction

In Chapter I of this Report it has been shown that modern biotechnology is opening up innumerable exciting new possibilities which may dramatically affect society over the next century. Some of these new developments are already in the arena of commercial exploitation, especially in the fields of health care and of agriculture. But the results of the interviews with nearly one hundred companies reported in Chapter II suggest that we are still only in the early stages of the full scale application of this revolutionary new technology. In this section we discuss the problems of diffusion of the technology through the economy as a whole and consider the probable time-scale of this diffusion process.

We shall make some comparisons with other pervasive technologies which have had very widespread economic consequences in the past, such as the introduction of electric power, and more recently computer technology and micro-electronics. It is always dangerous of course to make analogies of this kind and it is extremely important to take account of the *differences*, as well as the *similarities* between the various generic technologies which have so deeply transformed industrial societies over the past century.

However, although the unique features of modern biotechnology and its distinct areas of application must always be kept in mind, there are useful lessons which can be learnt from earlier waves of technical change which had very widespread economic and social consequences. It is evident, for example, that a transformation of the technologies in use in many sectors of the economy must lead on the one hand to large-scale investment in new types of plant and equipment and on the other hand to a change in the skill profile of the labour force. It must also lead to changes in company organisation and industrial structure, as is also evident from the results of the survey reported in Chapter II.

We know from the past experience of the introduction and diffusion of such revolutionary new technologies that these changes in capital stock, in the skill profile and in organisational structures of industry cannot possibly take place in a short period. They are a matter of decades rather than years or months. The recognition of the relatively long time scale involved in diffusion is extremely important as it can avert two dangers which might otherwise have adverse policy consequences.

First is the danger of "technological super optimism", which tends to ignore the hard economic realities of relative costs, profitability and size and consumer acceptance of entirely new products. Second is the danger of "technological conservatism" which fails to recognise the enormous long-term potential of generic technologies for the ultimate development of an entirely new range of products and services. The first can lead to serious under-estimation of the time-scale involved in diffusion processes; the second to equally serious errors of under-estimation of the potential of long-term transformation. We shall illustrate these points from the history of computer technology and electric power and we shall then consider the specific features of the new biotechnology and its diffusion in the economic and social system[9].

2. The analogy of the electronic computer: microelectronics and the "automatic factory"

We shall first take the example of electronic computer technology to illustrate the general problem of estimating the probable scale and timing. We take this example because (at least until the advent of biotechnology) it is probably the best known example of a pervasive technology in the second half of the twentieth century. Moreover it is one which is well documented and which is generally agreed to be of extraordinary importance for all OECD Member countries.

With the first application of the electronic computer during and just after the Second World War it was realised that this new technology had an enormous potential for transforming industrial processes, office systems and records and communication systems. However, opinions differed sharply on the probable time scale and extent of these developments. Some scientists and engineers anticipated very rapid and large-scale applications with immense social consequences, (including large-scale unemployment) already in the 1950s and 1960s. On the other hand it is well established that such a well informed industrial leader as Thomas J. Watson (Senior) did not believe, even in the early 1950s that there would be any big commercial market for electronic computers[10]:

> "He felt that the one SSEC machine which was on display at IBM's New York offices could solve all the scientific problems in the world involving scientific calculations'. He saw no commercial possibilities. This view, moreover, persisted even though some private firms that were potential users of computers – the major life insurance companies, telecommunications providers, aircraft manufacturers and others were reasonably informed about the emerging technology. A broad business need was not apparent".

It was not until the Korean War that IBM was persuaded to undertake production of a small batch of electronic computers and even then it was only with a change of management that they entered the commercial market. As against this conservative view of a very limited market for computers, imaginative scientists like Norbert Wiener (1949) envisaged a huge scale of applications and forecast large-scale unemployment as a result.

A much more balanced view of the probable time-scale and social consequences of the diffusion of the electronic computer was taken by John Diebold (1952)[11], one of the most authoritative and imaginative consultants in this field. In his book *The Advent of the Automatic Factory* he showed remarkable foresight and depth of understanding of the problems involved. Whilst recognising the enormous potential of the electronic computer for the transformation of all industrial and office processes, he saw quite clearly that this would be a matter of several decades and not of a few years. Indeed most of the "factory automation", which is today described as "FMS" (Flexible Manufacturing Systems) or "CIM" (Computer Integrated Manufacturing) did not show a really rapid take-off until the 1980s, even though most of the technical innovations which come under this heading were clearly foreseen by John Diebold in 1952. The "automation" of the 1950s was really a kind of advanced mechanisation (mainly in the automobile industry) rather than computerisation.

Diebold stressed several reasons for believing that the diffusion process would be much slower than many computer enthusiasts imagined at that time. The most important were these:

i) True computerised "automation" would involve the redesign of all industrial processes and products. It would be quite impossible to achieve this in a short period.

The simple availability of computers was only the first step. An enormous amount of R&D, design and new investment in machinery and instruments would be needed in every branch of industry.

ii) Such a process of redesign would affect *both* products and processes. Diebold gave examples to show that this could only occur if there was a change in the structure and organisation of firms, as well as in the attitude of management. This change would involve much closer integration of R&D, design, production engineering and marketing – a horizontal rather than a vertical flow of communication and information within firms.

iii) Not only would computer-based automation change the configuration and organisation of every factory, it would also involve a big change in the skill composition of the work-force. Diebold rejected the idea of mass unemployment arising from automation and also the idea of "deskilling" the work force. On the contrary he stressed the *new* skills that would be required, especially in design and maintenance, and saw automation as a means of overcoming the fragmentation and dehumanisation of work. But he also saw that it would take a long time before the new skills were available and people were retrained.

iv) Diebold recognised the importance of the *economic* aspects of diffusion. Computers would diffuse not only because they were technically advantageous. They had also to be *cheap.* It was only with the advent of micro-electronics in the 1960s and the micro-processor in the early 1970s that computerisation took off in small and medium-sized firms (SMEs), as well as in large firms, and in batch production as well as in flow process industries such as chemicals. Moreover, (and this is the most important point when we are looking at economy-wide effects) computer technology could only realise its potential outside a few "leading-edge" industries when computerised *systems* became relatively cheap and accessible.

Events since 1952 have fully confirmed Diebold's analysis. Even though the computer industry itself was growing at an extremely rapid rate for the next thirty years it took a whole series of complementary radical and incremental innovations such as Computer-Aided Design (CAD), Computer Numerical Control (CNC), Large Scale Integration (LSI) and big developments in software engineering and process instrumentation before computer-based automation could diffuse to most industrial and service sectors. Even now the economic advantages are by no means always clear-cut and there are often considerable teething problems when firms attempt to introduce FMS, or other forms of computerisation.

3. The analogy with electric power

A similarly long time scale was necessary for the diffusion of electric power and its innumerable applications, from the time of its first appearance in the 1880s, and for similar reasons. Not only did it take two or three decades before generating and transmission systems made the new energy source universally available in the industrialised countries; it took even longer to redesign machinery and equipment in other industries to take advantage of electricity, and to make the necessary skills available.

Warren Devine[12] has given an illuminating account of the debates which took place at the end of the 19th Century and the early part of this century on the implications of electric power for the future of factory processes.

"Replacing a steam engine with one or more electric motors, leaving the power distribution system unchanged, appears to have been the usual juxtaposition of a new technology upon the framework of an old one (...). Shaft and belt power distribution systems were in place, and manufacturers were familiar with their problems. Turning line shafts with motors was an improvement that required modifying only the front end of the system (...). As long as the electric motors were simply used in place of steam engines to turn long line shafts, the shortcomings of mechanical power distribution systems remained."

It was not until after 1900 that manufacturers generally began to realise that the indirect benefits of using unit electric drives were far greater than the direct energy saving benefits. Unit drive gave far greater flexibility in factory layout, as machines were no longer placed in line with shafts, making possible big capital savings in floor space. For example, the US

Table 4

CHRONOLOGY OF ELECTRIFICATION OF INDUSTRY[1]

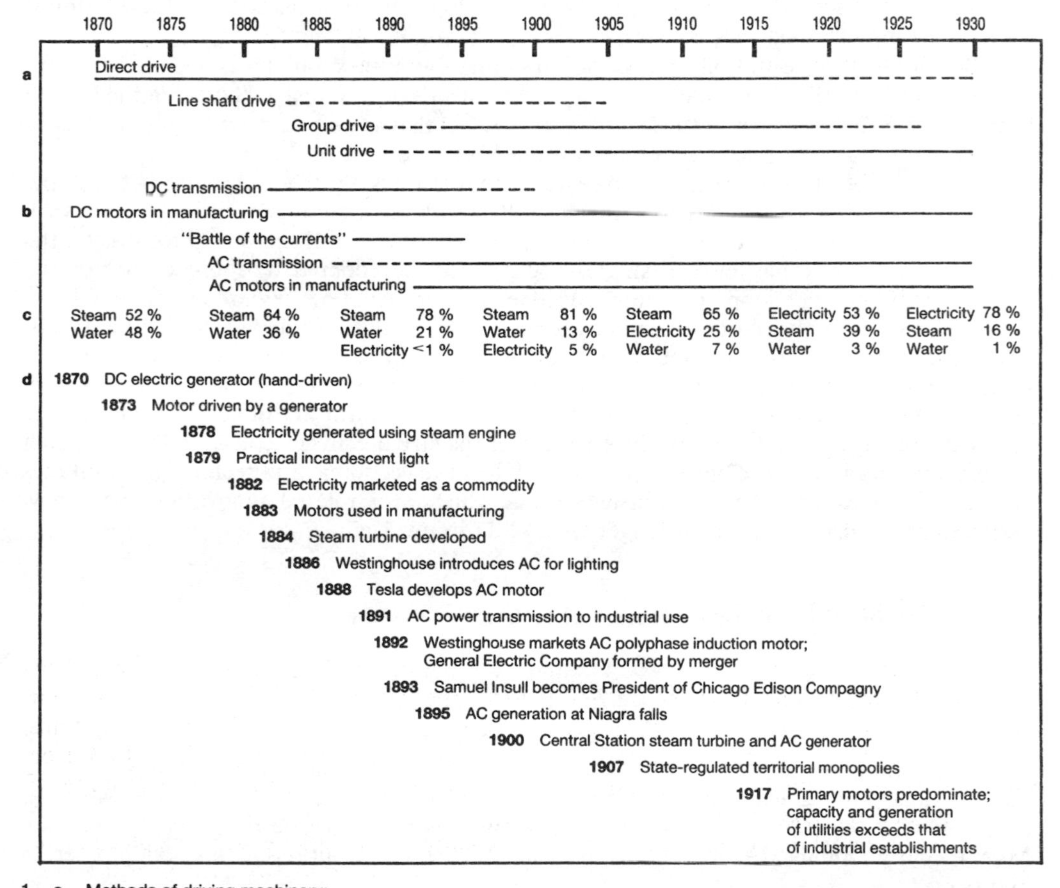

1. a. Methods of driving machinery;
   b. Rise of alternating current;
   c. Share of power for mechanical drive provided by steam, water and electricity;
   d. Key technical and entreprenurial developments.

Source: Devine, W.: From Shafts to Wires: A Historical Perspective, *Journal of Economic History*, Vol. 43, pp. 347-373.

Government Printing Office was able to add 40 presses in the same floor space. Unit drive meant that trolleys and overhead cranes could be used on a large scale, unobstructed by shafts, countershafts and belts. Portable power tools increased even further the flexibility and adaptability of production systems. Factories could be made much cleaner and lighter, which was very important in industries such as textiles and printing, both for working conditions and for product quality and process efficiency. Production capacity could be expanded much more easily.

The full expansionary benefits of electric power to the economy depended, therefore, not only on a few key innovations in the 1880s, but on the development of a new paradigm or production and design philosophy. This involved the redesign of machine tools and much other production equipment. It also involved the relocation of many plants and industries, based on the new freedom conferred by electric power transmission and local generating capacity. Finally, the revolution affected not only capital goods but a whole range of consumer goods, as a series of radical innovations led to the universal availability of a wide range of electric domestic appliances going far beyond the original domestic lighting systems of the 1880s. Ultimately, therefore, the impetus to economic development from electricity affected almost the entire range of goods and services.

But this complex diffusion process took about half a century and it was not actually until the 1920s that electricity overtook steam as the main source of industrial power in the United States (Table 4). It was not until the 1950s and 1960s that widespread ownership of electric consumer durables became the norm in Europe and Japan.

From these historical analogies it is evident that there is a major difference between the diffusion process for a single product and the diffusion process for a generic technology with numerous potential applications in a variety of different industrial sectors. Once it is on the market in an acceptable form, a single product may be adopted by more than half the population of potential adopters within a decade. This occurred for example in many OECD countries for a variety of consumer durables such as television, in the 1950s and 1960s. Mansfield's studies[13] also showed that this rate of adoption, occurred for some types of industrial and transport equipment, such as the diesel locomotive and the continuous strip mill. However there are other case studies of both agricultural and industrial innovations (such as Metcalfe's 1970 study of the diffusion of the size box in the cotton industry) which show many "laggards" and non-adopters even when the economic and technical advantages are clear[14]. When we come to consider whole clusters of related innovations with new generations of products, a much longer time scale in involved.

4. The new biotechnology as a change of technology system

a) *General criteria for major economic effects of new technologies*

Most diffusion research in the post-war period has concentrated on the diffusion of individual products and processes, and on incremental types of innovation. Schumpeter[15] was almost alone among leading 20th Century economists in looking at "creative waves of destruction" – the effect of major new technologies as they pervade the economic system. More recently a number of economists have made further contributions to this Schumpeterian approach.

Nelson and Winter[16] used the expression "generalised natural trajectories" to describe cumulative clusters of innovations, as for example, those associated with electric power. Dosi[17] used the expression "technological paradigm" by analogy with Kuhn's[18] scientific paradigms.

In these terms "incremental innovations" within an established paradigm may be compared with Kuhn's "normal science". Carlota Perez[19] has developed the concept of "techno-economic paradigms" to describe those changes in technology which pervade the entire economy and provide the new "common sense" for a whole generation of engineers and managers.

Clearly biotechnology is already a new paradigm in Dosi's sense and a new "natural trajectory" in Nelson and Winter's sense for the development of products and processes. Whether it is such an important trajectory that it will ultimately come to affect management decision-making in most branches of the economy remains an open question. The new biotechnology has undoubtedly led to enormous excitment in the research community and many new companies were established with venture capital to pursue R&D. This "research explosion" was without parallel. But the pervasiveness of a new trajectory or technology system depends on the range of *profitable* opportunities for exploitation. Until recently, despite its undoubted importance for the future, biotechnology has led to *profitable* innovations in only a relatively small number of applications in a few sectors and in a few countries.

In Schumpeter's model, the profits realised by innovators are the decisive impulse to surges of growth, acting as a signal to the swarms of imitators. But this "swarming" behaviour, generating a great deal of new investment and employment, depends on falling costs of adoption and very clear-cut advantages and/or competitive pressures. Later on, of course, after a period of profitable fast growth, profitability may decline. Schumpeter stressed that changing profit expectations during the growth of an industry are a major determinant for the sigmoidal pattern of growth. As new capacity is expanded, at some point (varying with the industries in question), growth will begin to slow down. Exceptionally, this process of maturation may take only a few years, but more typically it will take several decades and sometimes longer still. Biotechnology is a very long way from this mature stage and the main interest is in when it will enter the "swarming" phase and on what scale.

For a new technological system to have major effects on the economy as a whole it should satisfy the following conditions:

i) *A new range of products accompanied by an improvement in the technical characteristics of many products and processes*, in terms of improved reliability, new properties, better quality, accuracy, speed or other performance characteristics. This leads to the opening up of many new markets, with high and rapid growth potential and the rise of new industries, based on these products.

ii) *A reduction in costs of many products and services.* In some areas this may be an order of magnitude reduction; in others, much less. But it provides another essential condition for Schumpeterian "swarming", i.e. widespread perceived opportunities for new profitable investment. The major revolutions such as electric power and computing were both labour-saving *and* capital-saving, but also offered a reduction in the cost of other major inputs, such as energy.

iii) *Social and political acceptability.* Although many economists and technologists tend to think narrowly in terms of the first two characteristics, this third criterion is also important. Whereas the first two advantages are fairly quickly perceived, there may be long delays in *social* acceptance of revolutionary new technologies, especially in areas of application far removed from the initial introduction. Legislative, educational and regulatory changes may be involved, as well as changes in management and labour attitudes and procedures. Changes in taste, especially in sensitive areas such as food and drink, are often unpredictable.

iv) *Environmental acceptability.* This may be regarded as a subset of iii) above, but, especially in recent times, it has become important in its own right. It is of particular significance in relation to the *Limits to Growth* debate and the debate over nuclear power. It finds expression in the development of a regulatory framework of safety legislation, and procedural norms which accompany the diffusion of any major technology. Particularly hazardous technologies or those which are extremely expensive to control are severely handicapped, even if they do have some economic and technical advantages, as in the case of nuclear power.

v) *Pervasive effects throughout the economic system.* Some new technologies, as for example the float-glass process, have revolutionary effects and are socially acceptable, but are confined in their range of applications to one or a very few branches of the economy. It follows from 1 and 2 above that for a new technology to be capable of affecting the behaviour of the economy as a whole, in the Perez sense, it must clearly have effects on technical change and investment decisions in many or all important sectors.

b) *The case of new biotechnology*

Using these five criteria it is relatively easy to see why nuclear technology does not qualify as a change of techno-economic paradigm since it fails on almost every one of them. Electric power or the micro-electronic, computer-based information technology by contrast satisfy all five criteria.

Clearly, from the evidence in Chapters I and II of this report, the new biotechnology is likely to satisfy the first of the above criteria. It is beginning to give rise to a range of new products and processes in health care, in medical diagnostics, in water treatment, in veterinary applications, in agriculture and forestry, in the food industry, in services and in mineral extraction and processing. New varieties of plants with drought-resistant, pest-resistant and other specifically engineered attributes are one of the most active areas with enormous potential. This is confirmed for example by the structural change in the activity of chemical firms entering the seed and bulb industries on a large scale. The future potential is even greater, extending to a broad range of chemical and food products and processes and perhaps ultimately to an even wider spectrum. If the hopes of "bio-chips" are realised they could extend to the whole range of microelectronics.

When we come to the second criterion the picture is less clear-cut. Perhaps the best analogy is with the first two generations of electronic computers before the advent of the integrated circuit and the microprocessor. At this stage, in the 1950s, computers certainly found cost-reducing applications in such areas as pay-roll and invoicing, but the cost of computers was still relatively high so that the range of adoption was limited. In the case of the new biotechnology there are indications that rDNA derived and other new processes will be less costly than traditional manufacturing processes in certain health-care and agricultural sectors (Chapter VII, 2*d*), but it is also evident that in other important areas, such as animal feeding stuffs (Chapter VI, 5*b*), mineral extraction, energy and chemical feed stocks, the lack of cost competitiveness has slowed down or prevented more widespread application. This is an important limitation on the speed of diffusion in key industrial sectors.

However, these limitations may be overcome as a result of further research and development, or as a result of rising costs and diminishing supplies of non-renewable materials or both. In his analysis of the "Economic Potential of Biotechnologies" Rehm[20] pointed out that very much further research and development was needed not only in the field of chemical feedstocks but also in relation to biomass, metal leaching and oil recovery. Chapter I, 2*d*/*e* has shown that his point is still valid. Relative prices and estimates of probable profitability have not yet led to a major expansion of these fields of R&D.

In the NAS Report on *New Frontiers in Biotechnology*, Cooney[21] pointed out in 1984 that:

> "Products from the biochemical process industry (...) take advantage of the same economies of scale experienced in commodity chemicals production (...). Most biochemicals are made using inexpensive raw materials, such as sugar, and they offer good potential value added. The profit margins depend on the efficiency in transforming these raw materials into products. It is this biochemical problem that needs to be translated into a biochemical process. At this point one begins to see the need for integrating improved conversion yields, better metabolic pathways and new reactor mechanisms. This requires integrating biochemistry, microbiology and chemical process technology."

However, this integration of disciplines and skills is by no means easy to achieve as it requires new forms of organisation and structure in firms (and in universities). It is a problem comparable to that identified by Diebold in the case of factory automation. Diebold realised that an enormous amount of design and development work was necessary for each specific application of computers to industrial processes and that the skills were often not available for this work. Nor were firms organised to achieve results. In the case of biotechnology similar points are valid but the extent of re-design may be less far reaching. Postgate[22] may be right when he says that one major disadvantage of biotechnological processes need not prevent them from becoming cost effective (i.e. that the product usually has to be concentrated from relatively dilute solutions). He is right too that they have the advantage of not requiring high temperatures and pressures. However, the experience so far with the scaling up of biotechnological processes to meet the requirements for large scale production of *bulk commodities* is not encouraging in relation to comparative costs.

Costs remain high and it is an open question whether biotechnological processes will replace the present processes for bulk chemicals in the next 20 years. Applications in the copper industry appear more promising, as Warhurst has shown[23].

Links upstream to more fundamental research have been a central feature of the new biotechnology and exceptionally important for chemical and drug firms[24]. This will continue to be extremely important in relation to most new products. At present, the dependance of biotechnology on information technology is considerable. "Super-computers" and advanced information systems are essential for advanced research, development and design work in molecular biology. Biosensors also demand integration with electronic technology. The two technologies are likely to become increasingly interdependent in future generations of "intelligent" computer systems and process control systems (Chapter I, 3). But again the present experience of "5th Generation" computers and "artificial intelligence" suggest that these developments will also extend well into the next century and are surrounded by great uncertainty.

The discussion so far has concentrated mainly on the problems of process technology in relation to the potential large scale future applications of biotechnology outside the present rather limited area.

This is because the economy-wide effects of the new biotechnology depend upon the resolution of these problems (our second criterion). It is already clear that biotechnology is having effects in the pharmaceutical industry, medical care and agriculture. Whether these effects extend to the whole of the chemical industry, oil recovery, the energy industries, the food industry and ultimately an even wider range of manufacturing and service industries, will depend upon the progress of research, development and design over the next 10-20 years. This in turn relates to social and organisational problems in the "national system of innovation" – the management and scale of R&D, the interfaces between different parts of the system, the availability of skills, the encouragement (or lack of it) to the experimental application of new processes and so forth. Finally, the incentives to conduct R&D and to introduce new processes depend on the development of relative costs in alternative processes.

Whether or not the new biotechnology becomes a new "techno-economic paradigm" dominating future economic development in the next century, depends also upon whether it satisfies other criteria. So far we have discussed almost exclusively the first two criteria relating to technical and economic performance.

The third and fourth criteria are dealt with more comprehensively in the following Chapter (IV, 2) on public acceptance. Suffice it to say here that continued acceptance cannot be taken for granted, although the initial public debate and the regulatory mechanisms already introduced offer a favourable basis for diffusion in most countries. Various retarding factors in such sensitive areas as agriculture, food and public health are however present.

Clearly, new social and institutional problems will arise as the scale of application extends. These could be very great, involving for example, the restructuring of agricultural and health services on a worldwide scale as Chapter V will demonstrate in more detail. Kristensen[25] has suggested that there may even be a "role reversal" between the present group of industrialised countries and the present group of underdeveloped countries. Because of their land area, demographic pressures and environmentalist pressures as well as because of new technological developments, the present industrial (mainly OECD) countries may become the main suppliers of agricultural products (indeed Kristensen suggests this is already happening, even without biotechnology; see also Chapter II, 4*c* and Chapter VI). Some of the present Third World countries on the other hand, especially the NICs, may become the main centres of manufacturing production and exports during the next century.

This speculative "futurology" is intended *not* as a precise forecast of future events, which is in any case impossible, but to illustrate the type and magnitude of the structural changes and the social and institutional adjustments which may occur as biotechnology begins to have really widespread effects. Big changes in the internal structure of agricultural production within each country are also probable. One small example of this may illustrate the point. The UK is now exporting date palms on a significant scale to the Middle East. This business has been pioneered by what was once a small horticultural enterprise on traditional lines but is now a medium-sized firm with R&D facilities and hundreds of employees. Classical (and neo-classical) trade theory would probably have considered the idea of the UK being an exporter of date palms as absurd on grounds of comparative advantage. But new technologies change many parameters and this certainly applies to biotechnology.

c) *The pervasiveness of new biotechnology*

Following this brief discussion on the four criteria so far considered, we may now turn to the final (fifth) criterion to be taken into account in assessing the macro-economic consequences of the new biotechnology: pervasiveness.

The new biotechnology is clearly more pervasive than more narrowly focussed technologies, such as nuclear power. It has already found applications in primary industries (agriculture, forestry and mining), in secondary industries (chemicals, drugs, food) and in tertiary industries (health care, education, research, advisory services). However, a comparison between Chapters I and II clearly shows that the *actual* range of applications is still far narrower than the potential, and the "known potential" (a concept which may be compared with "proven reserves" in the oil industry) is still much narrower than in the case of computer applications.

In fact, biotechnology has often been compared to information technology, whose influence can be felt in all economic sectors. However, it is necessary to emphasize some fundamental differences at least for the time being. First of all, the fact that biotechnology operates through living organisms (or parts thereof) limits the field of activity to materials that can be biologically manipulated. Numerous industrial sectors would then be excluded from the *direct* influence of biotechnology (e.g. the metal and steel industry, mechanical industry, telecommunications and so on), although an indirect or mediated influence cannot be excluded.

Information technology, on the other hand, operates through substitution or change of a given production factor (i.e. labour) and has been able to penetrate almost all products and processes of human activity.

There is no process that has not been or cannot be modified by the use of information technology. From this feature derives the functional pervasiveness of information technology: informatics and telecommunications are employed not only by technical personnel (e.g. production, engineering, R&D), but also by non technical personnel (e.g. administration, financing, marketing, sales,). This does presently not hold true for biotechnology, which, in this respect, is more akin to chemical technology.

In the longer term, a linking of biological and information technologies might materialise in specific devices (biochips, neurocomputers, biorobotics, see Chapter I) endowed with much higher capacities for storage and processing. The merging of information technology with the power of the new biotechnology would give the latter all the pervasiveness-aspects of information technologies and would influence human activity in ways which are presently difficult to imagine.

As we have seen, the future path of diffusion is still surrounded by considerable uncertainties, associated both with future technological developments and with socio-political developments. For this reason, Figure 1 illustrates simplistically three possible "scenarios".

Scenario 1 would represent an accelerated diffusion of the new biotechnology into many new industrial sectors and applications, including the rapid development of many new industrial processes as well as products. It would represent a more rapid advance than that which occurred historically with earlier waves of new technology. For reasons which have been discussed (time and scale of R&D, education, training, capital investment, social and structural change) this scenario seems highly improbable. Only rapid changes in relative costs, prices and profitability might induce such a development.

Scenario 4 would represent a much slower rate of diffusion than now seems likely from the identifiable potential of the technology. It would mean not only that "bio-chips" but also many other of the hoped for applications of biotechnology in the chemical, food, agricultural, energy, medical, environmental, mineral and other industries remain still in the future even after another 20 years of R&D. This also seems a rather unlikely scenario in the light of Chapters I and II and from all that is known about the current pace of R&D activity and the technological potential.

Something between Scenario 2 and Scenario 3 therefore is identified as the most probable, with biotechnology beginning to be *a major basis* for new investment and the growth of the economy in the second or third decade of the next century.

Of course, the level of aggregation of this Scenario is very high. A more precise analysis would have to break down biotechnology into various fields of application in order to assess both promoting and retarding diffusion factors in each of them. Diffusion rates vary greatly

Figure 1. **A SIMPLIFIED ILLUSTRATIVE REPRESENTATION OF THE DIFFUSION OF "MEGA-TECHNOLOGIES"**

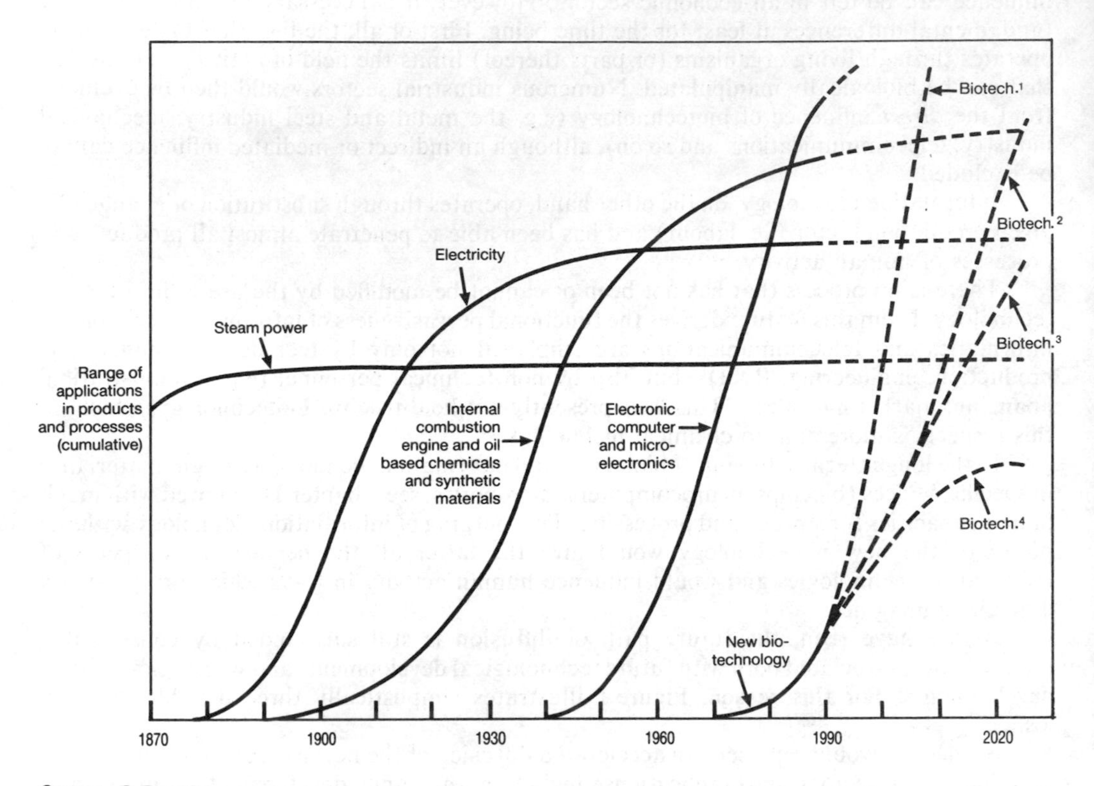

Source: C. Freeman.

between sectors. Health applications of biotechnology are likely to diffuse much faster than agricultural applications, and within the health sector, diagnostic products are diffusing faster than therapeutic products. Past experience in agriculture indicates that new technologies often take 10-20 years before they are adopted, but these rules may not always apply to the new innovations coming from biotechnology. If the introduction of an agricultural innovation has clear economic advantages, and if there is a fast flow of information, no supply bottlenecks and no regulatory delays, diffusion times in agriculture could be shorter than in the past. The new biotechnology appears to satisfy at least the first three of these four conditions. But even then, social and structural adjustments have to accompany and support technical change.

The fact that the new biotechnology is not likely to become the predominant technology for most industries and services in this century should be no cause for surprise. Nor does it mean that it is unimportant for economic growth and international competitiveness. On the contrary, it is clear that already it will be at the heart of a rapidly growing cluster of new industries and an essential element of competitive survival in an increasing number of established industries. Moreover, one reason for the intense research interest is that the unexpected can always happen in such a fast-moving area.

Finally, it is essential to remember that the discussion so far has been mainly in terms of the *quantitative* contribution of the new biotechnology to economic performance. The *qualitative* and social changes associated with its diffusion are likely to be far more important and cannot be measured in such categories as GNP or industrial and agricultural production as the Introduction has indicated. Its effects on health care are already profound and its influence on the entire culture and social fabric of OECD countries may be even more profound. These aspects of biotechnology raise fundamental philosophical and ethical problems which are beyond the scope of this section. Suffice it to mention here the extraordinary implications of "genetic fingerprinting" for crime detection, for personal medical records and prognosis and for the insurance industry. Although the economic effects of this new activity may be limited and it is likely to develop first as a "service" provided by specialised firms to government and other industries, it raises very fundamental issues of personal and social behaviour as well as legal issues. These *qualitative* implications of biotechnology are likely to be one of the biggest challenges over the next decade.

IV. CONDITIONS OF DIFFUSION

1. Introduction

The conditions of the diffusion of new technologies are less well known than the conditions of their development and production. Research on diffusion has perhaps begun later than research on innovation, or has been less thorough. More importantly, the diffusion process is much more complex than the innovation process, as it involves many more actors beyond the technical and economic ones. While it has been possible, as the preceding Chapter has shown, to draw up a list of generally valid principles for the diffusion of new technologies, it is more difficult to establish comprehensive lists of conditions for specific technologies, particularly biotechnology.

Governments can play, and have already played a major role in the diffusion of biotechnology, particularly through regulatory, but also through taxation and other policies. Policies aimed at sectors which are the main consumers of biotechnology products, particularly agricultural, health and environment policies, have affected the diffusion of biotechnology and will continue to do so.

The diffusion of *information technologies* was greatly accelerated by defense procurement for semi-conductors and computers. This has led to the question as to whether biotechnology could benefit from similar government strategies. The answer to this cannot be uniform, and may vary between sectors and between countries. It is clear, however, that many biotechnology products respond to major social needs (health, environment) and address public, government or government-regulated markets (health, agriculture, food). Thus, social needs, backed in some cases by public procurement, in others by legal or regulatory conditions, could play a role in the diffusion of biotechnology, a role which might, to some extent, be comparable to that played by past defense procurement in the diffusion of information technologies.

Government policies and safety considerations have been extensively discussed in other contexts[26]. This Chapter will focus on two conditions for the successful diffusion of biotechnology which, although known from other technologies as well, have, from the very beginning, appeared as exceptionally important in biotechnoloy: public acceptance and patent protection. Contrary to the structural adjustment needs which will be discussed in Chapter V, and which are specific to individual sectors, public acceptance and patent protection are both relevant to all sectors and applications of biotechnology.

2. Public acceptance of biotechnology

a) *What makes new biotechnology different?*

Public acceptance of new biotechnology, or public confidence in it, has emerged as a central factor – for some, the single most important one – in the diffusion of this technology.

We have seen (Chapter III) that the successful diffusion of new technologies is conditioned by a number of factors, amongst which are societal and environmental acceptability. The microelectronics based information revolution satisfied the criteria of societal and environmental acceptability, whereas nuclear power encountered increasing difficulties, which largely explains why its diffusion has comme to a halt in many countries.

Does the ongoing debate, particularly on safety issues and public acceptance, indicate the conditions under which biotechnology will satisfy the criteria of societal and environmental acceptability? Does it indicate where present trends may lead?

No other technology of the 20th century has human and social impacts as pervasive as biotechnology. In this sense, biotechnology is clearly not comparable with microelectronics which has few direct environmental or health impacts. Some observers who have watched how the pressure of public opinion put an end to nuclear power developments, have asked whether certain applications of biotechnology will not encounter similar obstacles, and ultimately, a similar fate. If biotechnology, as a broad generic technology, is not comparable to microelectronics, it is even less comparable to nuclear power, and it is important to understand why. Three main characteristics separate biotechnology from the controversial new energy technology of the 20th century:

- Nuclear power consists of a small number of core technologies (reactor, fuel, waste disposal). These may, to some extent, be improved but they are all indispensable and so closely linked that the failure or non-acceptability of any of them would lead to the abandonment of the entire technology. Compared to such a one-track technology, biotechnology already has great versatility. It covers a wide range of technologies derived from, and relevant to every form of life, from micro-organisms to man. Thus, holding up one development of this multitrack technology, does not put a stop to "biotechnology"; on the contrary, it may accelerate other developments. For example, temporary difficulties with the introduction of genetically modified micro-organisms into the environment are said to have accelerated plant research, and may help explain a recent increase in the number of plants submitted to regulatory authorities for approval; if there are fears, they are crystallising around invisible micro-organisms, not around visible and apparently controllable plants. Another example, inadequate patent protection for naturally occuring micro-organisms, the screening of which has been a basis of traditional and modern biotechnology, is encouraging the movement of industry into the new gene technologies which benefit from better legal protection;
- In the context of nuclear power generation technologies, the terms "health impacts" or "environment impacts" have clearly negative connotations. Societal acceptance of such technologies became more difficult when it appeared that they would have undesirable health and environment impacts. In contrast, biotechnologies have the potential to address hitherto unsolvable health and environment problems and to replace other potentially more harmful technologies. The goal to achieve such beneficial impacts on health and environment, together with the scientific interest in the new discoveries, explains much of the enthusiasm which accompanied the emergence of biotechnology. Thus, even if the concerns about possibly detrimental side-effects of some biotechnologies are justified – and whether they are is not discussed here – it has never been denied that numerous biotechnologies will effectively reduce risks and have beneficial health and environment impacts;
- A third feature of biotechnology, which could be decisive for its ultimate acceptance by society, is that the discussion on the risks and benefits began earlier than in the

case of any other 20th century technology, and before new products or processes existed. Contrary to what occurred in nuclear, information and other technologies, this debate, focussing largely on the new genetic engineering methods, has helped to shape the direction and rate of scientific and technical change from the very beginning, and before the first large industrial investments have been made.

Table 5 selects a number of key events in the United States where the main safety and regulatory discussions as well as many of the scientific and technical successes have occurred. The events listed in the two columns of Table 5 bear little or no direct relation to one another, but the lists show that, in the early stages, safety and regulation considerations have preceded scientific and technical developments by several years.

Table 5
A SELECTION OF KEY EVENTS IN BIOTECHNOLOGY

Research and Innovation		Safety and Regulations	
1977	First Successful Genetic Manipulation (Microbian Gene-Transfer, Cohen-Boyer)	1973	Letter of Professor Paul Berg to National Academy of Sciences warning of rDNA risks (US)
1978	First Successful Expression of Insulin in rDNA Micro-organisms (US)	1974	Asilomar Science Conference I decides Moratorium for rDNA-experiments (US)
1982	First Production of Human Insulin by Micro-organisms (US)	1975	Asilomar Science Conference II lifts Moratorium (US)
1982	First Gene-Transfer in Mammals ("Super-Mouse") (US)	1976	First rDNA-Safety guidelines (NIH) (US)
1982	First, unsuccessful Gene-Experiment in Man (Beta-Thalassemia) (Italy, Israel)	1985	Columbia District Court prohibits release of Micro-organisms into the Environment (US)
1986	First Release of rDNA Plants (US)	1986	US Court of Appeals overturns Columbia District Court decision (US)
1987	First Release of rDNA Micro-organisms into the Environment (US)	1986	First International Guidelines on rDNA-Safety Considerations, Paris, OECD, adopted by the OECD Council

b) *Risks and benefits: a global exercise in technology assessment*

The discussions on safety, risks and benefits, and public acceptance have lasted 15 years. For the sake of greater clarity, it has been proposed to distinguish between issues related to the "*acceptability*", and those related to the "*acceptance*" of biotechnology[27].

Acceptability derives from rational, scientific evaluation of biotechnological, mainly rDNA safety issues which, however, does not exclude rational dispute when different weight is given to external, social or economic criteria. *Acceptance* is the reaction of the public rooted in a larger number of motifs, including emotional ones. In cases where the public has shown concern about a technology, scientific acceptability is a necessary, but not sufficient condition of acceptance. When a gap between acceptability and acceptance appears, it will be a goal of public policies to attempt to close it.

Both acceptability and acceptance issues affect the diffusion of biotechnology. For example, a biotechnology company developed a new diagnostic test to detect the virus responsible for cervical cancer, but feared that the relevant drug authorities would not approve it (1987). This meant the company anticipated a problem with the acceptability of its innovation. A major food company developed a new rRNA-based yogurt that was not only economical, but satisfied all of its country's safety and health criteria, yet decided not to market it out of concern that journalists may quote this innovation in a misleading headline (1987). This company foresaw problems with the public acceptance of its products.

Scientific discussions on rDNA safety ("acceptability") have focused on four levels of application: research, industrial use, introduction into the environment, human gene therapy. Scientists are now much more optimistic than 15 years ago, particularly with regard to research and industrial applications. Differences between experts on these issues are rarer and more narrow than at any time during the last 15 years. There is a broadly based consensus that rDNA-techniques constitute a considerable improvement on conventional methods. It is true that organisms can be modified by conventional methods as well. However, there is a possibility that the intervention will affect not only the targeted gene, but also other genetic traits. rDNA techniques allow for tailor-made interventions affecting only the targeted gene.

There is also a broadly based consensus that any hazards arising from rDNA organisms are in general of the same nature as those of conventional organisms, and that any risks may be assessed generally in the same way as those associated with conventional organisms. A distinction is made between industrial and environmental application where, until 1988, it has not yet been possible to draw up precise safety criteria. Here a case-by-case approach is therefore recommended, until more experience has been gathered.

Experts believe that more research is necessary to develop risk assessment in biotechnology, particularly to analyse both the probability and the scale of conjectural accidents in comparison to other technologies. Thus, expert discussions on risks, safety and acceptability will probably continue as long as techniques relevant to biotechnology advance and multiply. However, the fact that these discussions began comparatively early has already had three major effets:

- Biotechnology laboratories and industry have been encouraged to choose low-risk micro-organisms;
- A public and institutional framework to review and if necessary supervise new biotechnology processes and/or products has been put in place in most, if not all, OECD countries with modern biotechnology laboratories and industries; and
- A strong movement towards international harmonisation of rDNA-safety criteria has begun within and beyond the OECD area, guided by OECD work. This harmonisation will facilitate the international diffusion of a technology which depends so heavily upon global markets.

The OECD publication on *Recombinant DNA Safety Considerations* (Paris, 1986) including the recommendation of the OECD Council to Member countries "concerning safety considerations for applications of recombinant DNA organisms in industry, agriculture and the environment", is considered a milestone on the road towards international harmonization. Already, most OECD countries have incorporated the OECD safety concepts into their national guidelines or regulations, and several non-member countries are taking them into account in developing their national safety policies in biotechnology.

In 1988, OECD work has been focussing increasingly on the safety evaluation of organisms to be introduced into the environment, as a growing number of such organisms are being considered for field trials. It is expected that present work will lead to a more detailled set of internationally accepted guidelines on biotechnology safety.

During the 1970s, "disenchantment" with technology in various OECD countries stimulated demands for a "technology assessment" function which was expected to move from limited cost-benefit analyses to a more general and permanent policy tool. The proponents justified their demand with the need to provide early warning, to prevent negative effects, to guide regulatory processes and, more ambitiously, to forecast and anticipate complex technological developments[28].

Those demands no longer dominate the technology discussion in OECD countries. However, almost unnoticed, the biotechnology policies of the last five years have moved in directions which would seem to satisfy many of them.

Compared to the scientific "acceptability" of biotechnology, it is much more difficult to assess and summarise "acceptance", because the public has understood the term biotechnology in various ways, and also because there are large international differences in public acceptance. These differences are deeply rooted in national traditions related to food, medicine, and health, which explains why they are changing only very slowly and why different countries do not seem to strongly influence each other.

In contrast to the scientific assessment of rDNA safety, there has been little or no international convergence of public attitudes towards biotechnology, neither in a favourable nor unfavourable direction. An opinion survey conducted in 1979 indicated wide variations between popular attitudes in European countries towards genetic research and R&D on synthetic food[29]. For example, 61 per cent of the interviewed people in Denmark thought genetic research to be "unacceptibly risky", while the corresponding figure for Italy was 22 per cent.

Not enough is known about the diverse social and cultural roots of attitudes towards biotechnologies, and why they vary so widely. Understanding public attitudes is important for policy formulation and perhaps even for predicting future trends. In fact, Denmark has subsequently (1988) been in the forefront of legislative initiatives constraining genetic engineering, while of all EEC countries, Italy has given least effect to the 1982 EEC Council Recommendation on the registration of rDNA research. Unfortunately, surveys of public attitudes are infrequent and often not comparable between countries. Regular and internationally comparable surveys of professional quality would be very helpful to policy makers and to industry.

Even if, more recently, some political efforts have been made in Europe to influence public opinion across national borders against biotechnology, it is rather revealing that the most prominent American opponent of biotechnology[30] on an information campaign in the United Kingdom in 1987, was barely noticed and apparently failed to get his point of view across.

c) *Government and industry responses*

While national differences may be difficult to interprete, it has in general not been too difficult to understand *what* the public's main concerns are. The following are most often mentioned:

- Ethical concerns about genetic modifications, in general or more particularly in humans;
- Safety concerns about health, and about the introduction of modified organisms into the environment;

- Concerns about the alleged, radical novelty of biotechnology, or about its alleged unpredictability or irreversibility; and
- Concerns about negative employment impacts.

However, these concerns are often found mixed-up with issues of health and life which have no direct link with biotechnology (e.g. *in vitro* fertilization).

The question as to *what* the main concerns are, however, is compounded by the problem of *who* exactly should be addressed to have those concerns alleviated. There are mainly two opinion trends on this problem which has often been debated. Some believe in the need to approach the public at large, the laymen, as directly as possible, without relying on intermediaries, or the media. Others consider this to be ineffective and propose to have the educational effort directed at those who shape public opinion, the politicians and the journalists, or at target groups which have influence and a high degree of credibility, e.g. teachers and doctors.

Governments, parliaments and, in some countries, industry have paid exceptional attention to public acceptance and public fears. Some reviews of public opinion have been conducted as indicated above[31], and numerous, often extensive debates have taken place in national parliaments, in some cases with results that *could* constrain rDNA applications, at least temporarily (European Parliament, Parliaments of Denmark and Germany).

Obviously, there can be no unified response to public concerns and no single recipe to gain public acceptance. The single most persistent response of governments to cope with the confusion between biotechnology and medical ethics in the public's mind, has been to keep debates and committees on human genetics or medical ethics strictly separate from those which review gene technologies for scientific and commercial applications.

Apart from this, it can be said that public authorities and industry have shown a remarkable degree of innovativeness, developing a wide range of responses to address public concerns. Amongst the main responses, some of which have also been used in combination, are:

- To "demystify" biotechnology by improving information for the public. This information does not necessarily emphasize increasingly complex scientific details, but rather the achievements, the breadth, continuity and antiquity of biotechnology (a technology with which mankind has been familiar for thousands of years, and from which it has derived substantial benefits). It could also be demonstrated that rDNA methods can provide greater safety than traditional methods. A large array of communication techniques have already been used: information campaigns, television programmes, tapes, books, industry visits, etc.;
- To find new ways to let the public participate in safety assessment. One case is the United Kingdom's Advisory Committee for Genetic Manipulation (ACGM) where trade union representatives and members of the public sit side by side with industry and government representatives. Another case is the United States Department of Agriculture's Agricultural Biotechnology Research Advisory Committee (ABRAC), which includes representatives of academia, industry, government and the public, including an ethicist and an ecologist. The chief purpose of these bodies is to achieve greater transparency in decision making, and the chief method is direct involvement of lay people, although it must be said that, in other countries, the participation of lay people has not always improved public acceptance;
- To acknowlege, in contrast to the preceding two strategies, that public information on biotechnology has sometimes been too extensive and hence counter-productive, because over-information on complex issues creates "cognitive stress". This would

explain why in some instances, public anxieties seem to have increased rather than decreased. What people need is more trust in biotechnology risk management, in the credibility of those who inform them, and in the willingness of governments and industry to abandon projects when risks become more important than benefits; and

- To abandon the general term biotechnology, and to replace it with the precise term of each individual technique or application which may be discussed. "Biotechnology" covers too diverse a range of apparently unconnected activities, some of which have been safely in use for centuries. Precision would spare the activities of the latter category from being dragged into the safety debate.

To anticipate future public reactions to biotechnology and their effects on diffusion is not easy. Various unpredictable events can influence public opinion:

- During discussions of new safety issues, disagreements between experts can emerge again. Although this would be a normal development in an ongoing debate between scientists of so many different disciplines, such disagreements could be amplified and misunderstood in public opinion;
- Public concern about accidents in other sectors, particularly chemical, could spill over into negative reactions against industrial technology in general, including biotechnology, even if the latter has, paradoxically, the potential to replace chemicals by safer biological components. Also, a wide diffusion of biotechnology, particularly in the Third World, could increase the possibility of accidents which may have repercussions on public opinion in other countries; and
- The social and economic adjustments required by the diffusion of biotechnology, may be difficult for, and resisted by certain social and professional categories. For example, biotechnology could become an additional threat to jobs in agriculture which has already created concern (Chapters II, VII).

As in the past, the effects of such events, or of some of them, on the diffusion of biotechnology might be relatively short-lived. More importantly, negative effects might be compensated by growing public demands for products and solutions which only biotechnology can deliver. The fear of AIDS, and the enormous public pressures put on academia, industry and governments to find rDNA-derived tests, vaccines and cures, offer a striking example of a large biotechnology pull-market created by public demand. Through its wide range of applications and impacts on man and society, biotechnology has the potential to generate large oscillations in public opinion, in both positive and negative directions.

If at the present moment, concerns about risks seem to outweigh the public appreciation of benefits, the obvious reason is that large benefits are expected from new biotechnology products of which there are still very few, whereas the risks are associated with already-existing rDNA technologies. It can reasonably be expected that an increasing number of products arriving on the health and environment markets will influence the public discussion in the coming years, in favour of biotechnology, particularly if governments and industry remain alert to public concerns, and are innovative in their responses.

The practical demonstration that biotechnology means products which can be very helpful to people will greatly enhance the chances of the technology to be accepted by the public.

Terminology is a major unresolved problem in communication with the public. The scientific language of biotechnology is a closed book to most people; the available training in the life sciences at all school levels, often does not prepare laymen to understand the terms or to appreciate the discoveries and opportunities of biotechnology.

3. Patent protection in biotechnology

Modern biotechnology has depended, from its beginning, upon an appropriate legal framework. This is the first case in history where the law has come to play a dominant role in the very emergence of a new technology. Among various legal issues, those related to patenting have remained a reason of concern for governments and industry alike.

The "Chakrabarty decision" of 1980, when the United States' Supreme Court ruled that a man-made, oil-eating micro-organism developed for industry was patentable under United States' law, has freed the discussion on patent protection in biotechnology from doubts as far as micro-organisms are concerned. The principle of patenting industrially useful micro-organisms is now widely accepted throughout the OECD area.

The crucial importance of this and other court decisions to grant patents to new biotechnology inventions was, in its time, hailed as a breakthrough in the history of modern biotechnology. In many OECD countries, law makers and judges have recognised that the new biotechnology, often derived from genetic modification, represented a great novelty due to human ingenuity, and a departure from past practice, and hence deserved a change or reinterpretation of law or legal practice, which is not easily granted. Their forward-looking understanding removed in time what could have become a great hurdle to the development and diffusion of biotechnology.

Thus, while the need to reassure a worried public has led policy-makers to emphasize the old roots and safe traditions of biotechnology, the need to improve patent protection has also led to greater emphasis on the scientific and technical novelty of the new inventions. The two positions, if not carried too far, are not necessarily contradictory as they address different biotechnology questions.

In fact, several other patent problem areas have not yet found solutions satisfactory to the inventor in biotechnology. In no other field of technology do national laws vary on so many points and diverge so widely as they do in biotechnology. It appears that United States law and Japanese law are, on the whole, more open and flexible towards the new developments in biotechnology than are the laws of other OECD countries.

Biotechnology is a particularly good example of the patent questions raised by rapid scientific and technological change. Change leads to problems of legal adaptation, particularly when the law is embedded in international treaties which can be changed only with the approval of a large majority of its signatories. The latest important international legal conventions relevant to biotechnology (International Convention for the Protection of New Varieties of Plants 1961, Strasbourg Convention 1963, European Patent Convention 1973) were discussed and ratified before the spread of the new biotechnological inventions, especially those of genetic engineering and therefore, do not take them into account.

A number of reform proposals have been made: those on the "grace period" and on plant protection deserve particular attention.

A *grace period* would allow scientists to submit, within a certain time limit, a patent claim on an invention, even if they have already disclosed it in scientific publications or in any other way that would contravene the "novelty" requirement of patent law. Grace periods exist in a few national laws (United States, Japan, Canada). These make patent law more compatible with the habits of the academic scientists who want to publish but would forfeit patent rights were they to do so. There is a widespread conviction in industry and universities that the importance of the academic contribution to the development of biotechnology will make the introduction of an internationally recognised grace period of six months (if not a year) more and more necessary. However, this proposal has met with opposition because, amongst other reasons, it would require a modification of the European Patent Convention.

Proposals related to *plant protection* have also led to controversy. In most countries, new plant varieties fall into the domain of special Plant Variety Rights. They have been applied to plant products obtained by traditional breeding methods which generally cannot be protected by patents. Except for the United States, the law of many countries, and specifically the European Patent Convention, excludes all new plant varieties from patent protection. However, new genetic engineering methods are increasingly being applied to plants and will lead to new plant varieties. Such methods may be the subject of written, scientific description and repeatable as required by patent law. The question arises, therefore, why these methods and the products thereof should not be protectable by patents. The US Patent and Trademark Office has, in 1985, granted a first plant patent for a genetically modified maize plant. Industry, which carries out much of the plant genetic R&D, considers patent protection a much better and commercially more attractive alternative than plant variety rights. In this case, opposition to reform in Europe and elsewhere has arisen from plant breeders and agricultural lobbies.

A comparable prohibition exists in the European Patent Convention against the *patenting of new animal varieties* and new technical breeding methods, whereas US law is, again, more open. In 1988, the US Patent and Trademark Office has granted a first animal patent to the inventor of a genetically modified mouse to be used in cancer research. The increasing economic importance of transgenic animals, and the fact that their genetic modification can be scientifically described and repeated, might stimulate a re-examination of this legal prohibition in Europe.

There are other components of patent protection particular to biotechnology (claims to *naturally occuring micro-organisms*, *release conditions* for micro-organisms which have been deposited for patenting purposes) which have been found to be unsatisfactory. Also, among general patenting conditions, there is one which inventors in biotechnology would particularly like to see improved: the *lifetime* of patents. It has been a regular complaint that the often protracted patenting procedures for biotechnology inventions are seriously eroding the normal 17-year or 20-year lifespan of patents.

The arrival of the first new biotechnology products on the market has recently (1987) been reflected by an increasing number of patent controversies. The interpretation of the law by the courts could affect the economic diffusion of biotechnology to a considerable extent. The question, which *scope of patent claim* would be reasonable in relation to a given invention ("inventive step"), has, at the end of 1987, become one of the disputed issues. Could the entire field of genetic manipulation with all its applications be pre-empted by one inventor or one research group because it claims to have made the first fundamental invention in this context? If very broad patent claims can be sustained before the courts, the costs of new biotechnology may increase, which could have a dampening effect on diffusion.

Experts believe that the large-scale diffusion of new biotechnology will critically depend upon internationally more harmonised patent protection. Discussions to this end have begun at national and international levels, encouraged by the OECD[32] and co-ordinated by the World Intellectual Property Organisation (WIPO) in Geneva as well as other bodies. However, traditional principles of law are not easily abandoned, and reform proposals take time before they are accepted and implemented. In the past, international patent law has changed very slowly, perhaps once in a generation. Several signs indicate now that we may be at the beginning of a major change of the patent system. If this change is to include a modification of international treaties, such as the European Patent Convention, it is not likely to be completed before the end of this century – still in time for the large-scale diffusion of biotechnology, but not a moment too early.

V. THE NEED FOR STRUCTURAL CHANGE

The preceeding chapters discussed the complex problematic related to the diffusion of new biotechnology and mentioned that structural change and system transformation are prerequisites of a pervasive diffusion throughout the economy.

This chapter focuses on some of these changes, emphasizing particularly those in the health and agriculture sectors, which will be first affected by new technology. Assessing structural changes inevitably implies an element of speculation as biotechnology is still in its infancy in terms of new commercial products or processes.

Beyond those two areas, there are many others where biological techniques find important applications, from mining industries to energy production. Chapter III has indicated, however, that here the biological processes are not yet able to undermine the dominant technologies, but rather represent an added tool to resolve specific problems.

1. Impacts on products and processes

A number of biotechnology developments will have profound impacts on processes and products. Structural changes in economy and society, are likely to follow from these technical changes. Some of them are already detectable in the progress of biotechnology and will be briefly sketched: a new emphasis on diagnosis and prevention; an increase in the specificity and effectiveness of new products; a reduction in the intensity of use of energy and materials which has been called "dematerialisation"; and increasing compatibility of technology with the environment and with natural resources.

a) *A new emphasis on diagnosis and prevention*

The first two chapters indicated that new biotechnological methods of diagnosis and prevention are expected to multiply in the fields of health, agriculture and the environment. Already some of the greatest successes of new biotechnology are tied to the commercial introduction of the growing number of immunodiagnostic tests based on monoclonal antibodies, biosensors and gene probes[33]. These new devices will allow for an extension of hitherto limited physical and chemical measurements, to a wide number of organic molecules, as well as for potential control and regulation of complex systems in the human body, in animals, plants, the environment and in industrial processes. The new tests are rapid and highly specific. The range of applications could be extremely wide: clinical controls and online monitoring of patients; *in vitro* and *in vivo* diagnostics of man and animals; monitoring of the effectiveness of drugs; quality control of food, air, water and soil through detection of pollution agents; criminal investigation; industrial purification of biological products and online control of biological processes (see also Chapter II, 4).

In another field of health care, new vaccines for humans and animals presently under study, will significantly increase the number of diseases which can be dealt with through effective preventive measures rather than with more costly therapeutic treatment[34].

What has been said for health care is valid also in other sectors. In agriculture and forestry, trends towards early detection of infestation or plant diseases through different techniques are emerging as well as preventive trends. This can be seen in both the mass production of clones with desired characteristics and the transfer of new traits to plants using the techniques of recombinant DNA. The aim of these efforts is the production of superior plants with qualities fitting the specific need of the eventual user or the production of plants highly resistant to disease, pests, climatic and environmental conditions (Chapter I, 2*b*).

The new emphasis on diagnosis and prevention is by no means restricted to biotechnology: most modern technology is directed towards early detection of critical symptoms and hindrance of degradation. Parallels can be found in rather distant sectors, such as that of materials; here also one sees an increase in the use of methods of preventing corrosion (cathodic protection, for example), of early diagnosis (ultrasound), and a development of new materials with superior resistance to corrosion.

b) *Dematerialisation*

Several authors[35] have shown that in industrialised societies, materials and energy inputs into the economy tend to diminish at least in relative terms: this can be measured at the macroeconomic level for most OECD countries with the reduction of intensity of use of materials and energy (expressed as the relationship between energy or materials demand and the GDP, in kcal/$ or kg materials/$). Such reductions are linked with the displacement of the production mix towards "light" industry and services, which in general are low hardware-intensive sectors and with the impact of modern technology which acts towards efficiency in the use of resources, optimisation processes, increasing effectiveness, lightness and specificity of materials, miniaturisation of objects, waste reduction, and so on.

New biotechnology operates in the same direction and will contribute to reduce energy and materials needs per unit of GDP, thus accelerating the movement of industrialised societes "beyond the era of materials[36]". Examples of this include: production of chemicals through enhanced fermentation, enzymatic processes, tissue culture or cell culture; the production and use of more efficient biopesticides, which are specific and biodegradable; highly efficient separation of molecules (industrial purification) by monoclonal antibodies; the insertion of a pest resistance function into the genome of plants to reduce the use of pesticides; the substitution of sugar by new compounds with dramatically superior sweetening power: 1g. of thaumatine is equivalent to 2kg. of sugar (Chapter I, 2*c*).

Dematerialisation may also have structural aspects. One is a shift in the utilisation of natural resources from rare to more abundant raw materials. Modern technology allows man to generate new materials and energies from easily available and cheap sources. It is possible to produce electricity without high-energy content fossil fuels, but by photovoltaic conversion of solar energy, the least expensive and most abundant energy source; the key material of the electronic age has been and to the largest extent, still is silicon, easily obtainable from sand; superior thermal and mechanical performances are associated with ceramic materials, most of which are based on silica and alumina, also easily obtainable from a variety of raw materials.

Biotechnology contributes in very similar ways to dematerialisation. The first practical results of genetic engineering have been obtained by using modified micro-organisms to mass produce in fermenters already known pharmaceuticals which, up to now, have been obtainable only in small quantities (growth hormone) and/or at high costs through extraction from human and animal organs (insulin): this means replacement of scarce and highly valuable "raw materials" by common fermentation substrates. Even in the cases where advances in chemical synthesis could replace biotechnology processes (insulin), the trend towards replacement of scarce by abundant raw materials will not be reversed, but rather reinforced.

Similar trends can be found in agriculture: comparative advantages related to soil and climate are increasingly being reduced (Chapter VI, 1), while artificial systems substitute for the natural environment (greenhouse), for soil (fertilizers) and for natural organisms (clones, transgenic plants). In an extreme case, soil will no longer be necessary to grow vegetables and land will simply become the physical support of an industrial factory: this is already the case in cell cultures for the production of secondary metabolites used as fine chemicals for cosmetics and pharmaceuticals[37], as well as in phytobioreactors for the production of microalgae.

Quantitative and structural dematerialisation of economic activities induced by biotechnology may lead to better environmental compatibility of new products and processes, as the amount of materials to be mobilized decreases and rare resources are preserved. However, other biotechnologies may also create waste problems of a new and different nature.

c) *Towards a rationalisation of the innovation process*

The exceptional potential of biotechnology relies on sound and rigorous scientific knowledge in numerous fields. The methodologies employed in the development of new products and processes increase rationality, while the contribution of pure empiricism is diminishing. This phenomenon is fully apparent in the radical change which has occurred in the last decade in the approach to the development of new drugs. The method of screening of large number of molecules (i.e. serendipity) has been to a great extent abandoned in favour of a large number research aimed at understanding the mechanism of various diseases. With this knowledge it becomes possible to target a suitable molecule to act upon those mechanisms. The change in the paradigm of pharmacological research has simplified the process of innovation and has made it more rational. This change, made possible by biotechnological research instruments and products, has profoundly affected the pharmaceutical industry. A similar evolution towards rationalisation of the innovation process in industry is expected to emerge in other sectors where new biotechnology may apply, from agrochemical to food industries.

2. The health care system

The "wonder drug" – highly specific, effective, without side effects – has been a dream which advances in biotechnology will hopefully bring nearer to reality. These advances offer the possibility to collect more precisely, and on a wider scale information about biological and biochemical aspects of the organism, and about the mechanism of action of drugs, in order to arrive at a *more specific drug design, with fewer secondary effects*. An example of this can be seen in the possibilities offered by the identification, isolation and production of new factors; e.g. proteins with different functions in the human organism, which can be used as drugs themselves (tissue plasminogen activator as thrombolitic agent), or as instruments to develop new drugs (for receptor identification). Also, monoclonal antibodies can be used as *ultraspecific drug vectors* against specific tissue antigens (targeting).

These advances will not come about without problems; the complexity of research certainly has not diminished; the costs of new drug developments are high and the time required long; and new problems of risk assessment and experimentation continually arise.

The slowness with which new therapeutic drugs become available means that structural change in medical practice is not imminent and not immediately required as a result of *therapeutic* innovation. This may change in the future as a result of the introduction of medicines specific to *individual patients*, such as personalised cancer therapy (LAK therapy), gene therapy, anti-idiotype monoclonals for auto-immune diseases, and others.

a) *Mass diagnostics*

However, economic, social and institutional changes deriving from the wave of immunodiagnostic tests based on monoclonal antibodies and gene probes will be faster and deeper. These tests arrive on the market at regular intervals, as their development requires relatively minor costs, time and risks, at least compared to therapeutic drugs.

Further structural impacts may result from the introduction of numerous new vaccines still at the experimental phase. These new vaccines based on rDNA-technology will be safer than traditional vaccines, both for users and producers. But their increasing numbers, and the resulting opportunities and needs for large-scale vaccinations, will call for a different organisation of health services than the one which has been adequate for the 4 or 5 vaccinations which have been common in industrialised countries until now.

The diagnostics revolution will cover a broad spectrum, including prenatal diagnosis, early diagnosis of the onset of diseases, and the monitoring of degenerative diseases. The new tests have many interesting characteristics: rapidity, specificity, facility of use, a wide spectrum of applications and great sensitivity to small quantities of test material, which means a dramatic reduction in the quantities of material needed (blood, urine, cells, etc.). They are appropriate for periodic check-ups of individuals, and their use extends to the *general population*. This is a powerful diagnostic instrument which could bring about an important reduction in the damage caused by disease. However, to make possible a mass diffusion of the new technologies, biological innovation should be accompanied by innovation in other areas, such as: sampling automation (blood, urine, cells), automation of instrumentation in order to quickly determine many parameters with little organic material from a large number of individuals; the development of expert systems for the processing and interpretation of analytical data; telecommunications for the transmission of data from laboratories to data processing centers; use of software languages which allow for maximum confidentiality in the use of data banks which collect information on the state of health of citizens. The potential benefits for individuals are as significant as the potential impacts on health-care systems which are still overwhelmingly focussed on therapy, not to speak of the public health budgets which generally operate in very difficult financial conditions. In fact, the main driving force behind the establishment of data-banks in health care is the need to reduce costs, particularly labour costs. The introduction of mass diagnostics will greatly add to this force.

The availability of a large number of simple diagnostic kits will also favor the diffusion of *tests* performed at *home* or by the *doctor*. The task of the general practitioner could be facilitated by tests which will precisely determine many types of diseases, thereby permitting the formulation of a specific therapy without the laborious routine which requires that the patient first be referred to a specialist and then to an analysis laboratory. The course of the disease may also be followed precisely by the general doctor, using quantitative tests. The advantages both for the patient and for the efficiency of the health care system are obvious.

Moreover, for some biological parameters, there are already some simple "do-it-yourself" immunodiagnostic tests available. The best known example is the home pregnancy test. Here also, the change is important because the possibility of self-control of infective diseases or of biological parameters becomes available at the patient level.

It is true that home tests are, for the time being, still in their infancy. Can they be recommended in cases where medical counselling is of prime importance? In cases such as AIDS or cancer diagnosis, home tests may produce more problems than solutions. Governments will have to license them with care and may, for certain diseases, prefer doctors' tests to an indiscriminate use of home tests. A more complex and costly but better controlled type of "home test" is the online monitoring of patients through biosensors and communication systems which would allow for following-up patients at home.

It should also be mentioned that technological progress which substitutes laboratory by home tests, corresponds to an apparently growing social demand for self-diagnosis and self-therapy which opinion surveys have recently detected in industrialised countries.

The transfer of technology from the laboratory to the doctor (doctor's test) or to the private individual (home test) represents a great organisational and functional innovation. Since this breaks with traditional practice, obstacles from the health care sector, from conservative bureaucracies and from professional categories which have a vested interest in the existing system, are predictable.

b) *Necessary changes in public health care systems and in industry*

What are the implications of current technological evolution for health-care systems, health costs and the pharmaceutical industry?

The transition from laboratory to doctors' and home tests will require institutional and educational adjustments of various types, the most important of which are the retraining of doctors in clinical biology, and the education of patients in the use of these tests. Other adjustments are the modification of regulations concerning the carrying-out of biological analyses, and the reimbursement by public and private health funds of expenses sustained by doctors.

More generally, a shift of emphasis from therapy (hospitals and drugs) to mass-diffused diagnosis (kits) and prevention (vaccines) presupposes an organisational and cultural restructuration of national *health insurance and reimbursement systems*. It must be expected that, in many countries, such change will be difficult given the inflexibility and politisation of public health services and structures. The greatest risk to the diffusion of new biotechnology is the lack of comprehension of the far reaching nature of coming change and the defense of corporate interests. Also, the technical and organisational problems in the management of the necessary data systems are numerous. They presume a transformation of the professional status and culture of the involved labour force.

The *cost implications* of the public health changes brought about by new biotechnology are complex and controversial. It will take much more experience before they can be fully assessed. For the time being, earlier hopes that the move from therapy to diagnosis and prevention would lead to clear-cut reductions in public health costs, have given way to doubts and caution.

There can be no doubt that this move will greatly improve quality of life, and have substantial, indirect macroeconomic as well as private economic benefits due to the reduction in the number of working days lost through sick-leave, and the likely prolongation of useful working life. The direct financial consequences for public health budgets could be quite different according to whether they are seen in a short or a long time perspective, and whether the emphasis is on diagnosis or on prevention. Seen in isolation, early diagnosis of many, particularly infectious, diseases should lead to reductions in both private and public health costs. Public health administrators, however, are concerned that the long term net effect of a move towards diagnosis may *increase* rather than decrease public health costs, particularly if diagnosis is not followed by effective therapy. If it is, the public health cost implications could be more encouraging, and so should be the health cost effects of large-scale disease prevention through a growing number of vaccines. Of course, the general ageing of the population, and increasing health quality demands of the public might in the long term compensate for any health cost reductions brought about by new biotechnology.

Apart from diagnosis, the monitoring of therapy (particularly home therapy) will become more and more important, both because of new types of therapy, and because of the new diagnostic tools which make monitoring possible. This again implies institutional and organisational changes. In some countries, the structure of the health care system at present will make continuous monitoring of therapy financially impossible. In other countries, insurance companies now seem more ready to reimburse the expenditures of home treatment including monitoring at home.

The *pharmaceutical industry* has been for some time in a state of change: from product supplier (principally drugs) it is becoming an "industry of function" or a health-care industry (i.e. a supplier of a wide range of therapeutic products, diagnostics, auxiliary materials, equipment, machines, biomedical systems and technology). The scientific basis is more and more interdisciplinary; relying not only on chemistry, biology and medicine, but also on physics, electronics, computers, lasers, etc. These transformations have already brought with them problems of internal reorganisation.

In the future, the massive diffusion of diagnostics will probably move the current process of transformation further towards an extended multidisciplinary base. The diffusion of mass health screening through the population carries with it not only biological tests, but also the parallel development of automatised equipment and software for data processing. The trend is, therefore, towards more integration of biotechnology, microelectronics and telecommunications. Thus, in the sector of medical technology, the traditional pharmaceutical industries may find themselves competing with the electronics industries.

This is a new challenge to the innovativeness of the pharmaceutical industry; it may be noted in this context that it is not the pharmaceutical industry which has first developed the new diagnostic tools, but the small biotechnological start-up companies. The critical success factor in this new challenge is probably the capacity to run complex systems rather than biotechnological knowledge in itself; it is therefore possible that the electronics industry will find itself at an advantage in relation to the pharmaceutical industry.

The organisational and institutional implications of this evolution are far-reaching. Such developments offer private enterprises the possibility of extending their activity to services, directly running diagnostic centers and, in this way, substituting for the public health system. This emerging trend could spread, above all in countries where the public sector has great difficulty in adapting to technological change: governments would be obliged to turn to the organisational support of private industry.

3. Agro-industrial systems and biomass

In the agricultural sector, the new biological techniques may have a very dynamic power extending beyond agriculture itself, to ancillary activities (i.e. fertilizers, agrochemicals, machinery, seeds, services) as well as to other downstream activities concerned with food, biomass and the transformation of agricultural products, including forestry and lumber, fishery, aquaculture, animal husbandry, pulp and paper, leather and natural rubber. Moreover, the use of innovative products and processes in the above-mentioned industrial sectors can in turn determine substantial changes in equipment, with further impacts on yet other sectors. A discussion of changes in agribusiness related to new biotechnology necessarily involves the complex question of biomass utilisation (that is the constituent material of vegetable and animal organisms), as agricultural surplus-production in many OECD countries is causing great concern.

Impacts of biotechnology on agro-industrial systems can already be felt and Chapter III has mentioned that the diffusion-time of agricultural innovations is perhaps getting shorter. In various OECD countries, it has been noted that farmers are becoming less conservative, better trained and more open to technical change. Also, the strong industrial involvement in agricultural biotechnologies is an accelerating factor. The technology to genetically modify plants is becoming available now but it will take approximately until the mid-1990s before the farmers will have the new seeds. Their diffusion could, afterwards, be very fast if the four conditions mentioned in Chapter III are met (clear economic advantage, fast information flow, no supply bottleneck, no regulatory delay).

a) *Impacts on quantity*

The development of agrochemical technologies of fertilisation and crop protection, of irrigation, mechanisation and of genetic selection of plants and animals has contributed in the past to an extraordinary increase in agricultural production in all parts of the world. This development has permitted the production of adequate quantities of food and even production of quantities exceeding demand. It should be underlined that many of these well-known technologies (e.g. cereal hybrids, artificial fertilisation, embryo transfer) are still at the diffusion stage, so that agricultural productivity is destined to grow at significant rates in the years to come.

Also, the powerful tools made available by new biotechnology are directed towards traditional lines of agricultural research, which aim at increasing production and/or reducing production costs. In the animal sector, the direction is towards reducing losses from disease or towards increasing production. Similarly, research in the fields of plants and forestry has often quantitative goals, being directed towards crop protection and the mass production of clones in order to standardise plant species.

Research also focusses on transgenic plants, seeking results superior to those obtained through conventional breeding. The objectives are the production of species which grow rapidly and/or are resistant to disease, chemicals, insects or adverse weather or environmental conditions or to produce plants that can grow on land which is presently not exploitable.

There certainly are agricultural commodities where quantitative production increases would be much appreciated by OECD countries. For example, countries which have a trade deficit in soya or in lumber may welcome the potential of new biotechnology to increase the production of both.

However, the introduction of this great mass of new know-how will bring about a further increase in agricultural production on a broad front, because productivity will grow, new land may be put to use and by-products may find new applications. While increased efficiency of agricultural production and reduction of resource inputs must be encouraged, the net effect of the new know-how could be to exacerbate the problems of excess agricultural production in OECD countries as long as there is no fundamental change in the latters agricultural policies and systems. New solutions to the surplus problem must be found which do not rely on subsidies that place too great a financial burden on governments; for instance, the fermentation of large amounts of biomass to produce ethanol for use as gasoline additive, would arouse opposition because it requires at present oil prices, conspicuous government subsidies.

b) *Quality impacts and new uses of biomass*

Resistance to the introduction of new biotechnology into agriculture has been fuelled by concern about the quantity impacts of the new technology. It is increasingly evident that agricultural biotechnology should be directed more towards qualitative goals than quantitative production increases, and towards the development of novel industrial uses of biomass. It is also evident that the tools of modern biotechnology will be indispensable to achieve these goals.

Quality has several facets. First, there are market demands for food with better taste and aroma, at least in wealthy countries (Chapter II, 3*b*). This is one aspect of quality open to new biotechnology processes, although the improvement of flavours, etc., through genetic engineering is still a very difficult field.

A second quality aspect relates to safer food and food with fewer chemical residues from pesticides and other agro-chemicals. On both accounts improvements are desirable and will become possible, thanks to biotechnology. Food contamination has been a source of many health problems, and it is often difficult to detect its exact origin in the food chain. Biotechnology could help to solve this problem through new diagnostics and prevention. Also, some governments are likely to prohibit in the coming years, chemical residues in food products such as milk, which asks for new biological approaches to replace present agro-chemicals.

In a third more general sense, quality means greater specialisation and diversification of products in order to respond to specific demands. Products derived from agriculture or forestry, may, in many cases, better respond to differentiated needs, than synthetic or inorganic products. In particular, the growing need for biodegradability and environmental compatibility of products, may be better satisfied through transformation of biomass, e.g. a major field of interest is that of completely biodegradable surfactants obtained from fats and sugar. Furthermore, in some markets which are particularly sensitive to "natural" factors (cosmetics, dietetics, pharmaceuticals), agriculturally derived products may well enjoy a distinct advantage.

In the competition between natural and synthetic products, the former have an advantage when molecular structures are complex, when the application occurs in sectors sensitive to public attitudes favouring "natural products", and when biodegradability is a critical success factor. Of course, careful assessment will be required in every case, taking into consideration factors such as relative prices of raw materials (oil versus biomass), relative costs of processes and associated environmental impacts, reliability and quality of raw materials, and valorisation by the market of natural origin, as compared to synthetic origin of products.

The development of new, economically viable uses of agricultural products, particularly for industry, is a critical challenge of our time, and biotechnology greatly adds to the currently available chemical and physical processes to transform otherwise useless biomass. However, this is a very complex task. It includes the development of new products to absorb not only excess cereals (typical of temperate climates), but also other types of products from different climatic and geographic conditions. This strategy could also help countries to find new uses for marginal lands (hills and mountains) which are being abandoned whenever governments consider this to be an ecological or other risk.

Such measures will take time, and they cannot under any circumstances solve the *present* surplus problems of OECD countries. Also, they should not be expected to greatly affect the historic downward trend in agricultural employment of OECD countries. However, while the development of industrial uses for agricultural products is unlikely to reverse this downward trend, it might, in some instances, slow it down (Chapter VII, 2*b*).

c) *Agribusinness: convergence of agriculture, industry and services*

The proposed shift of emphasis in agriculture from quantity to quality, and from food surpluses to new industrial products, calls for structural transformation in agribusiness, and for changes in the interaction between agriculture and industry. Trends are already emerging towards a closer integration between agriculture, input industries and user sectors. For example, the transfer of the crop protection function from pesticides to biotechnologically modified seeds has structural implications for the concerned actors, as can be seen in the large-scale acquisition of seed companies by the agro-chemical industry. In the future, productive diversification and development of new crops can succeed only with extensive co-ordination between agriculture and users. Industrial systems offer agriculture an ever-growing range of services, from input (chemicals, seeds, technology, finance, insurance, etc.) to output (harvest, delivery, commercialization). The trend is towards an integration of input and output industries, which would operate as a global service to agriculture. The farmer, in turn, would change his activity almost into that of an industrial entrepreneur; he would delegate several functions to specialists whenever growing technological complexity requires specialised competence.

The interdependence between agriculture and industry is expected to grow with the rate at which agricultural activity absorbs the increasing contributions of technology, be it to services, know-how, or to product differentiation and sophistication.

In this framework of growing ties, the chemical and food industries, which have a "culture" related to biology and are therefore better able to understand the needs and potentials of agriculture, tend to assume an increasingly central role, to the point of co-ordinating input and output, thereby transforming agriculture into a global agribusiness.

Progress in biotechnology, added to a number of technical and organisational innovations, may speed up the change in the role of agriculture in our society; no longer a backward activity, limited to satisfying human food needs, it could become a complex and wide-ranging industry which will supply primary materials to other economic sectors.

Although the most dynamic actor in the transformation of agriculture is likely to be the industrial enterprise, governments and farmer lobbies will play equally critical roles. Governments will have a decisive influence on this transformation as their intervention in agriculture is presently at an all-time high, and as it is in their power to determine the rate of change in agriculture. Changes in traditional attitudes and policies which up to now have been principally concerned with the protection of commodity prices and other short-term problems, will be required in the coming years.

In addition, it will be necessary to convince the farmer lobbies, perhaps even before governments, of the need of transformation. In many countries, this will presently be less difficult than it would have been a few years ago.

VI. PROSPECTIVE IMPACTS ON INTERNATIONAL TRADE AND COMPETITIVENESS

As a recent OECD study on biotechnology[38] has shown, the enhancement of international competitiveness has been up to now the main and possibly the overriding goal of government policies to support biotechnology, in particular biotechnology R&D. The pursuit of such a goal requires an international perspective on developments in biotechnology, including their potential impacts on trade and competitiveness. Given the limited number of products being marketed (in particular in the area of new biotechnology) it is of course too early to assess most of these impacts or to evaluate the effects of national policies on competitiveness. This is why this chapter aims primarily at developing an analytical framework which may help governments: *i)* to set biotechnology in the context of present patterns of international trade; and *ii)* to decide the main thrust of policies for competitiveness. For the sake of greater clarity the chapter begins by some overall analytical considerations.

1. Trade and technological change: theory and historical experience

At any given period, the pattern of international trade is shaped by: *i)* the characteristics of the goods supplied and demanded in the international market; *ii)* the location of the main centers of supply, as shaped by productive capacity and levels of competitiveness; and *iii)* the location of markets, the size of which, in terms of effective (e.g. monetary) demand is determined by size of population and more important still, the level of *per capita* income. In a science and technology-intensive economy such as the one in which the OECD countries – and by extension and impact the world economy as well – have lived since the end of the Second World War, change in the nature and characteristics of traded goods is very largely the result of supply side initiatives taken by innovators and entrepreneurs. Today resource endowments and hence comparative advantage, have become less and less linked to climatic and geographical factors and more and more shaped by "man-made" *innovation* and *investment driven phenomena.* In turn the capacity to innovate, vested in the R&D, human resources and capital investment outlays made by countries and individual firms, is the central factor which generates incomes and markets.

This, of course, is why the innovation-intensive and high income countries which are members of OECD occupy the central role in world trade, representing both the major outlet for the exports of developing countries and of the centrally planned economies, and the area *within which* the *single largest part* of world trade takes place. OECD markets are the most important outlet for OECD supply and the terrain where the most intense competition between Member countries takes place (Table 6). After increasing fairly significantly as a result first of the 1973 and 1979 changes in the price of oil and second of the gradual

Table 6

THE OVERALL STRUCTURE OF WORLD EXPORTS IN 1985

Percentage

ALL GOODS

Origin / Destination	Developed countries	Developing countries	Centrally planned economies	World
Developed countries	49.8	13.1	3.4	66.3
Developing countries	15.3	6.1	1.5	22.9
Centrally planned economies	3.1	2.0	5.7	10.8
World	68.2	22.2	10.5	100.0

MANUFACTURED GOODS

Origin / Destination	Developed countries	Developing countries	Centrally planned economies	World
Developed countries	58.1	16.5	4.2	78.8
Developing countries	8.1	3.6	0.7	12.4
Centrally planned economies	1.6	1.6	5.6	8.8
World	67.8	21.7	10.5	100.0

Source: GATT, *Report on World Trade*, 1986-87, Tables A2 and A3.

emergence of the newly industrialised countries (NICs) as exporters of manufactured products (both final products and semi-processed goods and other intermediary inputs to production), the overall share of developing countries in world trade dropped again from 1981 onwards while the growth of their manufactured products exports faltered as from 1985 (Figure 2). A number of factors are obviously at work, including the well-known factors related to Third World debt and protectionism in OECD markets, but technological factors are also considered by experts to be playing a part in the tendency towards renewed concentration of trade within OECD. Biotechnology is likely to be one of the forces acting in this direction.

a) *Technology, competitiveness and comparative advantage*

In a dynamic, innovation-driven economy, comparative advantage is hard to distinguish from competitiveness as expressed itself by the price, quality and *novelty* of products. At certain periods, competitiveness is, in fact, strongly determined by the type of competition which Schumpeter (1947) defined as "competition from the new commodity, the new technology, the new source of supply, the new type of organization (the largest-scale unit of control for instance) – competition which commands a decisive cost or quality advantage and which strikes not at the margins of the profits and the outputs of the existing firms (and possibly even entire national economies) but at their foundations and their very lives[39]".

After having been totally ignored by mainstream neo-classical international trade economists for many decades, there has been initial recognition of this Schumpeterian approach by the leading trade economists. As stated by H.G. Johnson, "innovative capacity

Figure 2. **SHARE OF DEVELOPING AREAS IN WORLD MERCHANDISE TRADE 1970-85**

(Percentage shares)

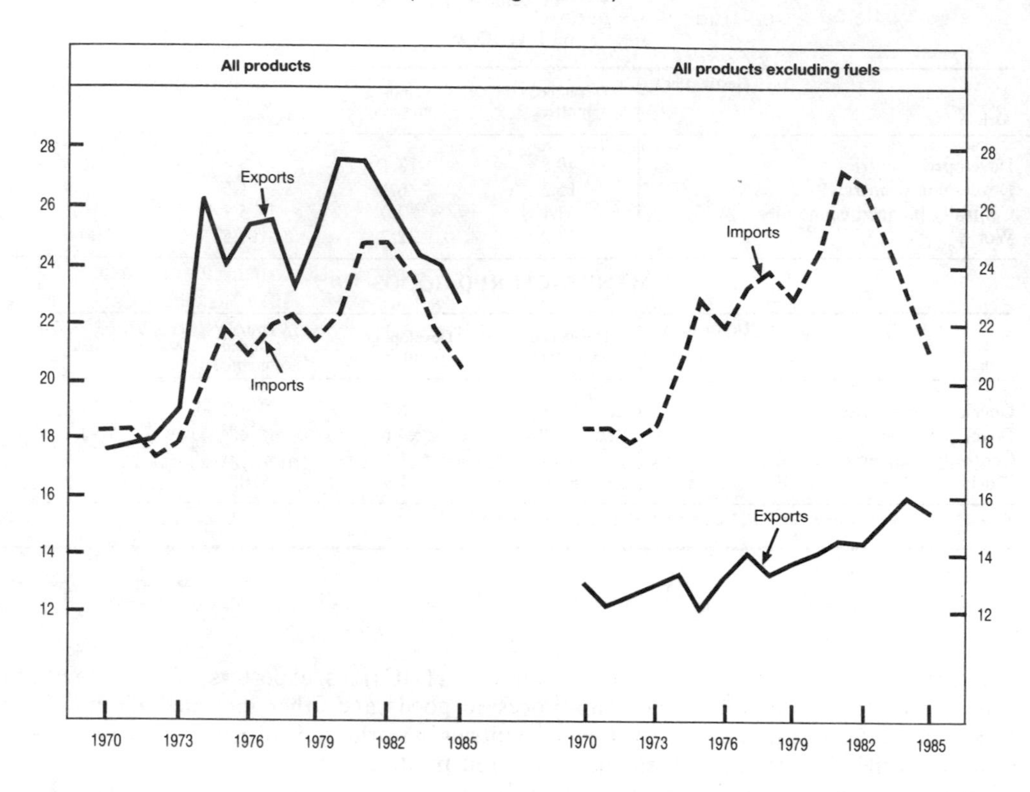

Source: GATT, *Report on World Trade,* 1986-87, Chart 1.1.

should be viewed as a basic source of difference in comparative advantage, and technological change as a chronic disturber of existing patterns of comparative advantage[40]". These "chronic disturber" effects of technology may occur in two main ways: *i)* by modifying trade flows as a result of the marketing either of totally new products or of new, better and differently sourced substitutes, whether they are material inputs to other industries or products for final consumption; and *ii)* by creating differences or gaps between countries, or suddenly widening gaps which were previously being reduced as a result of international transfers of technology.

A closer analysis of the trade modifying effects of technology shows that these can occur through four main channels or mechanisms:

i) The *creation of new trade* (either along existing overall international trade patterns or with a recognisable change in patterns) through the marketing through exports of *totally new products*. Stress is placed here on the words *through exports* to indicate that in the area of high technology (e.g. R&D intensive) industries, the worldwide marketing of new products may take place through delocated manufacture and the foreign operations of multinational enterprises (MNEs) and have little effect on trade flows *per se*;

ii) *Shifts in the structure of trade*, marked by the reduction and at some stage possibly the outright disappearance of particular trade flows, resulting from the *creation of entirely new substitutes* for previous products;

iii) *Shifts in the structure of trade*, again marked by the reduction of particular trade flows, resulting from the introduction of *new production processes* which change major factor proportions (e.g. capital/labour ratios) and reduce trade flows based on an abundant cheap labour type of comparative advantage;

iv) *Shifts in the structure of trade*, also involving a reduction in the level of trade flows, which *stem from the reduction of material inputs to production*, as a consequence of a number of parallel and/or related processes of *economisation* and *substitution*.

This fourth process which is of particular importance in modern biotechnology (see Chapter V, 1*b*), has been overlooked for a long time but has recently been studied fairly systematically by trade experts and international organisations in relation to the fall in developing countries' exports of primary metals and agricultural raw materials. The evidence has led UNCTAD to give its support to what a number of economists and technologists now designate as "a generally neglected dimension of long-term structural change which can be characterized as a '*dematerialisation' of production* that is, a shift in the composition of demand in industrialised countries away from the products of the more intensely raw material-consuming industries and a diminution in the intensity of raw material use in existing manufacturing industries[41]".

b) *Innovative capabilities, trade impacts and the ability to shape new technological paradigms*

If the distinction proposed by Freeman and Perez[42] between: *i) incremental* innovations; *ii) radical* innovations; *iii) technological revolutions*; and *iv)* historical transitions to *new techno-economic paradigms* is accepted, the three latter types of innovation may all be expected to have trade creating, substituting or displacing impacts. However even incremental innovations will influence the process of dematerialisation. In periods of technological revolution and historical change in basic paradigms the impacts on trade are likely to be accelerated, possibly with dramatic effects on the production and export of given products.

The extent and nature of such impacts on international trade patterns and country specialisations will depend on a number of factors: the scale and speed with which new output related to the radical or revolutionary technologies is marketed; the level of development, size, wide or narrow trade specialisation and industrial sophistication of national economies, and also of course on the overall economic climate in which the trade impacts take place.

The trajectories of new technologies[43] are rarely "natural" (e.g. in the sense of being commanded essentially by endogenous scientific and technological factors). Economic and social factors are of paramount importance in shaping trajectories and determining the way in which the new techno-industrial paradigms emerge[44]. The process is one of *selection* through the play of economic, political and social forces, at the basis of which lies of course a sufficient degree of *indigenous* scientific, technological and industrial capacity to be able to participate in this selection process.

Countries, industries and firms with strong scientific and technological capacities may be predicted to be on the initiating end of such processes or at least to have a reasonable hope of adjusting to them successfully, through investments in R&D, innovation and related shifts in industrial specialisation. As previous historical experience has shown, on the contrary "countries with weak scientific and technological capabilities will often be on the *receiving end of such processes* and may consequently be forced to bear the full brunt of adjustment through painful changes in exchange rates, employment and real incomes[45]".

c) *Technology, foreign direct investment and the process of globalisation*

The commercialisation of products through exports is only *one* of the several ways in which firms can exploit the temporary monopoly-advantages and firm-specific assets stemming from technological lead times and unique experience with new technologies. Such advantages are more and more often exploited through foreign direct investment and the international network of delocated production units based on transnational or multinational enterprises and corporations. A number of factors lie behind this development, in particular: *i)* the large range of factors which place a *premium* on delocated production inside foreign economies, *inter alia* non-tariff and reglementary barriers to trade (some of which may take the form of government regulations regarding health and safety) but also proximity to scientific and technical skills; and *ii)* the special imperatives of oligopolistic competition in industries where concentration has developed and where firms must be present on a fairly large scale inside the home markets of major competitors in order to wage competition successfully.

In industries where R&D costs are high, market niches for new products in the early stages of innovation small, and hence cash flows from innovation investment low in relation to R&D and start-up investment, these factors coupled with the imperatives of rapid commercialisation may often push firms to accentuate their multinational production and marketing strategies at the expense of exports. A subsidiary, generally considerably less advantageous form of recouping R&D costs and reaping benefits from innovation is through foreign licencing and/or the establishment of joint ventures and other interfirm technological and industrial co-operation agreements. This course is one which small innovative firms or else larger firms with low levels of multinationalisation may be forced to adopt because they do not possess the complete range of assets required to reap the profits stemming from their innovations.

Delocated production and the emergence of global competition on a world level in markets with a small number of competitors, have two major implications for the level and pattern of foreign trade and the international competitiveness of national economies: *i)* delocated production resulting from foreign investment will tend (as in pharmaceuticals, see below) to replace trade flows; *ii)* firms will choose their competitive strategies on the basis of considerations pertaining to world markets and the global relationships with their main competitors[46]. They will forego national or even regional (e.g. European) considerations when shaping their strategies, thus complicating the attempts made by governments to formulate and apply policies aimed at enhancing national or regional competitiveness.

2. Biotechnology, trade and competitiveness: some overall patterns

The concern by governments with future competitiveness *in* biotechnology, as well as with the effects of biotechnology *on* the *overall* competitiveness of domestic economies and *on* the scale and pace of technology driven structural adjustment, notably in agriculture, is well-founded and legitimate. Biotechnology has all the necessary requisites to usher in a set of technical and organisational paradigms in the area of health care, medicine and agriculture and will also offer new choices for the most efficient route for the production of many chemical products. Consequently biotechnology also has the requisites of the type of competition described above by Schumpeter; it will create trade, but it will also have strong trade displacing effects. At the moment, however, the impacts on trade and competitiveness are hard to measure and even *the trends difficult to define completely*. The outstanding feature of the situation *at present* is essentially the fact that *competition is being waged principally among OECD firms and countries, with pressures for structural adjustment mainly concentrated in specific areas of agriculture, but that trade impacts are being experienced principally by certain categories of developing countries*. However, should developments such as the modification of cereal grains (wheat) become a reality as predicted (see Chapter I), substantial impacts on agricultural exports from OECD countries would ensue.

a) *Biotechnology and competitiveness: novel features among OECD countries*

Other studies have already suggested that in the case of biotechnology a number of factors "preclude a traditional analysis of international competitiveness[47]". The first factor relates quite simply to the impossibility at present of measuring and comparing performances. Previous OECD work has shown that competitiveness is as much, and in new industries and technologies generally more, a question of quality and novelty than of price. In such industries indicators of relative cost (e.g. wage levels) and price simply do not represent meaningful proxies of competitiveness[48]. Competitiveness has to be measured through foreign trade data and derived calculations, e.g. shares in world exports, ratios of foreign penetration of domestic markets, at the most disaggregated level possible. In the case of biotechnology, and in particular new biotechnology, the level of production and sales simply do not permit any such measurement.

The second point made in the 1984 OTA report regarding competitiveness is of greater analytical interest and may be more durable in its influence on the way competitiveness among OECD countries will shape up in biotechnology. "Even with many more products on the market, a traditional competitive analysis might not be appropriate because an economic analysis of competitiveness usually addresses a specific industrial sector. The set of techniques that constitute biotechnology, however, are potentially applicable to many industrial sectors[49]." Even if biotechnology seems, for the time being, a less pervasive generic technology than microelectronics (see Chapter III), its range is, potentially, very wide. One aspect of the highly ubiquitous nature of biotechnology and the array of technique it calls on, is that it offers, in the long term, unique opportunities for creating totally new products, opening up totally *new markets* for which there exists *at present no competition* and so *no issue of competitiveness*.

A further novel feature of competitiveness in biotechnology is the fact that the contours of competition are shaped by previous processes of technological accumulation, industrial concentration and multinationalisation. Biotechnology has begun to grow not exclusively, but to a considerable extent, within the framework of already strongly concentrated and highly globalised industries (pharmaceuticals, chemicals, multinationalised food processing). In conditions where profitable market niches are still small and current cash flows insufficient to meet high R&D outlays, the pressure on firms to establish their strategies for the worldwide sourcing of scientific and technological resources and the global marketing of products is particularly strong. Firms may tend, perhaps more than in some other areas, to "go it their own way", without putting great expectations in national policies for competitiveness, which may explain why very many firms are apparently not impressed by their countries' policies (see above Chapter II).

These particular features have a number of implications for trade and competitiveness within OECD, *inter alia* the fact:

i) That *initially* at least firms and countries will probably, to a fairly significant extent, seek (as they are now in general doing) to use biotechnology as a way of *consolidating* and *enhancing* their "comparative advantage" (in pharmaceuticals, food processing, natural resources processing, agriculture, etc.) based on technological accumulation undertaken in earlier phases of industrial development;

ii) That there is less risk, in the longer term at least, for "picking the winner" strategies by firms as by governments to create trade tensions and trade issues, since they are less likely than in the case of commercial aircraft or automobiles for instance, to be concentrated on exactly the same products and to occur in slow growth markets. It is true that today numerous companies are pursuing the same few product developments principally in pharmaceuticals (Chapter II) and that the race of firms to be first to commercialise these products on a world-scale lies behind some major patent-disputes and behind the trend towards increasing secretiveness in R&D. However, a prospective analysis of competitiveness and its related issues must emphasize that the present situation is necessarily temporary, that the number of new products arriving on the market are expected to increase fast in the coming years, and that the range of *totally new*, or *old, but hitherto, not addressed* individual and collective needs (as in medicine, healthcare, waste disposal and environmental management) which could be catered for, is potentially so large that if these needs were to be recognised and met the market would be a very open and *large* one;

iii) That for an OECD country today, being "competitive" in biotechnology has still essentially the meaning of taking the *necessary steps to prepare the future*, to avoid being on the receiving end of the process of "creative destruction" (Schumpeter) and to participate as fully as it can in the scientific and industrial development of biotechnology and so in the new commercial opportunities it will offer; and

iv) That large firms possessing an advantage in biotechnology may tend to establish their own competitive strategies, without waiting for governments to act and without necessarily welcoming government plans.

b) *Biotechnology and trade substitution in raw materials*

Today, scientific and technological capacity in biotechnology is concentrated to a very high degree in the advanced industrialised countries of the OECD area. Consequently, it is these countries, on the basis of the economic opportunities and constraints perceived by their major industrial actors, which have been shaping the course of development in biotechnology and the "trajectories" along which it is progressing at present. Developing countries are only participating in the process *marginally*, even in cases where they have an indigenous capacity as in China, India or Brazil. As a result, at the moment, developing countries lie mainly on the receiving end of present clearly perceptable trade impacts.

These trade impacts are nearly all of a trade substituting character [types *ii)* and *iv)*]. As shown in Chapter IV, biotechnology is acting as a major agent in the process of "dematerialisation of production". Consequently it is quite certainly called upon to act as a further factor reducing the overall demand for primary products from developing countries. This process is clearly discernable since it has its origins in the advances of modern biotechnology even prior to recombinant DNA.

As has often been said biotechnology is not an industry *per se*, but a set of related technologies and scientific and technical methods, which encompass both modern fermentation and enzyme driven transformation processes, and the new technologies arising from recombinant DNA, protein engineering and cell fusion: in other words, they also encompass "second generation" or "modern" biotechnology, as distinct from the "third generation" or "new" biotechnology discussed above in Chapters I to III. While scientific and technological advances are blurring more and more the frontiers between the varying technologies and reducing the validity of a separation between the two "generations", a discussion of *trade impacts* must, for the time being, *retain the distinction.* Large marketed output and measurable trade impacts are associated primarily with the last phases of "second generation" or "modern" biotechnology and concern mainly agricultural products. They have already started to affect developing countries.

The analysis of actual and potential impacts of biotechnology on developing countries given below does not paint a very "optimistic" picture. This must be *qualified* by the attempts made to examine biotechnology in a "South-South", developing-country oriented perspective. UNESCO, FAO and UNIDO have in particular indicated how biotechnology could also help to meet Third World needs. This is particularly true of some of the worst health problems of the Third World (malaria, hepatitis and other infectious diseases, diabetes) which new biotechnology could help solve. In food, important research is also being undertaken by the International Rice Research Institute in Manila. However given the technological and industrial preponderance of OECD countries, and given the present assessment of firms regarding developing country markets as summarized in Chapter II, there is strong evidence that developing countries, notably those heavily engaged in agriculture, will bear the brunt of trade impacts for a long time to come.

3. Factors affecting future competitiveness among OECD countries

For reasons just explained the use of the term competitiveness in biotechnology is, for the time being at least, somewhat different from its use in the context of many other technologies and industries. This does not mean however that countries can overlook the scientific and technological *lead-times* in terms of the time required to train the appropriate scientific and technical manpower, build the appropriate R&D infrastructures and create the appropriate conditions for transferring the new technologies to industry.

While there is certainly still time for all OECD countries to participate effectively in the overall development of biotechnology, thus shaping the technology's specific trajectories and to prepare to be competitive in the production and commercialisation of biotechnology in given areas of relative excellence, the present situation requires action by countries without unnecessary delay. A wide range of factors may have an effect on competitiveness[50]. The following points to a few key issue areas.

a) *Technical manpower and scientific bottlenecks*

Biotechnology is a result of a process with particularly strong "technology-push" features. It is also at a stage where the pace of development and the relative competitiveness of countries continues to depend strongly on a number of factors relating directly to the *growth of the science base*, the overcoming of scientific bottlenecks and further *organisation of the science-industry interface*. This is *not an area where the policy issues can be couched mainly or exclusively in terms of technology diffusion.*

This is reflected in various studies of competitiveness-related issues in this area, which include recognition of education and infrastructure building priorities for government action. A recent study prepared for the US Department of Commerce[51] emphasizes that in Europe, and by inference to some extent also in Japan and even the United States, rapid growth of the knowledge base is likely to be limited by a number of *scientific and technical "choke points"* of which a list is proposed.

While financial and manpower constraints will impose priorities for countries as for firms, the *generic* nature of biotechnology and the early stage in its development, still argues in favour of fairly wide support of biotechnology R&D and against too narrow an approach to R&D priority-setting at this phase. This appears to be corroborated by the current performance of major chemical and pharmaceutical corporations in relation to biotechnology (both "modern" and "new"). At present the companies positioned most favourably are those which have put up the earliest and highest stakes for R&D and have funded research quite widely.

b) *Data banks and specialised software*

In biotechnology, the science base includes a very important component which is possibly called to play a wider role in growth and competitiveness than for many other technologies. This component relates to the role played by data banks, information-storing systems and the software needed to retrive and use such data in the course of research and/or production. In biotechnology, nucleic acid sequences, protein structures, MMR and 2D gel electrophoresis research are examples of research producing vast quantities of vital factual information of permanent value which exemplify the need for computer treatment rather than traditional paper information, manipulation and dissemination methods. The United States holds a commanding position with respect both to the size of its data bases and its accumulated experience in using them, and to the production of software designed for molecular biology and bioprocessing.

In terms of size, such data banks have no equivalents in other OECD countries. The European Nucleic Acid Data Base, the Nucleotide Sequence Data Library (operated by the European Molecular Biology Laboratory, EMBL), is extremely small in comparison, while Japan has only recently begun to build this type of facility.

In general, however, firms from Europe and Japan have tended up to now to rely strongly upon access to the US data bases for the information services they need. There is increasing doubt whether this approach remains realistic today given the growing commercial and strategic importance of scientific information and the steps which firms are increasingly taking to protect, not only their own firm specific knowledge, but also information emerging from the university and government research laboratories they are working with. This explains the priority the European Community has been giving to strengthening European data bases and improving the co-ordination between national systems.

c) *The industrial base and the research-industry interface*

However complex the support base for competitiveness may be in relation to a given technology and a given set of industries, ultimately competitiveness is dependent on the efficiency of the industrial base and the competitive capacity of the final link in the chain, namely firms. The fact that the new biotechnology is of particular importance in industrial sectors shaped by earlier processes of concentration and multinationalisation raises a number of important issues for competitiveness in a national or a regional (e.g. European) context. These are not necessarily peculiar to this sole area, but are probably raised more acutely than in many other industries. Two issues are of particular importance: the place of small firms in the industrial base (see Chapters II and V), and the possibilities offered to multinational enteprises (MNEs) of locating a part of their own firm-specific industrial and technological base outside their home country.

The "decentralised concentration" described in Chapter II places large firms in the position of being the principal form of industrial organisation through which the industry-research interface can be organised with a view to competitiveness[52]. A number of implications stem from this, notably the fact that the course of biotechnology's future technological trajectories are likely to be traced principally by the strategies and earlier lines of development of these firms.

Building the industrial base for national competitiveness in biotechnology around large well-established pharmaceutical, speciality chemicals or food-processing firms raises other issues for governments which derive from the multinational character of many of these firms and the global nature of their reach and strategies. Global reach and strategy by home-based multinational enterprises has important advantages as well as drawbacks. On the side of the advantages is the short term competitive asset the MNE brings to its country of origin on account of its capacity, through R&D contracts and research partnerships with foreign universities but also with small foreign biotechnology firms, and through production and marketing joint ventures with other foreign firms, to draw upon the resources located in a foreign science-base. Table 7 provides information concerning the number and nature of the international biotechnology agreements established between US firms (in particular the small new biotechnology firms) and foreign corporations of European and Japanese origins.

Table 7

INTERNATIONAL BIOTECHNOLOGY AGREEMENTS BETWEEN UNITED STATES AND NON-UNITED STATES FIRMS (1981 – 1st Quarter of 1986)

	Acquisition	Venture capital	Contract R&D	Joint R&D	License/ Production	Distrib./ Marketing	Joint Prod. Marketing	Total No. Contracts[1]
Belgium	1	1	1	3		1	1	9
Denmark		1	2	2	1	1	1	7
Finland			1		1	3		3
France	4	1	6	3	4	6	5	21
Germany	1	1	7	6	14	18	3	31
Italy		3	1	3	2	1	1	10
Netherlands	2	3		2		2	1	9
Norway		1						1
Spain		1		1			1	2
Sweden	3	3	5	4	5	6	3	19
Switzerland	1	2	11	4	5	6	5	26
United Kingdom	4	11	4	5	4	5	5	35
Total Europe	16	28	38	33	36	49	27	173
Japan	2	24	30	24	40	59	24	141

1. Some contracts may involve more than one type of agreement.

Source: Yuan, R.T., "An Overview of Biotechnological Transfer in our International Context", *Genetic Engineering News*, March 1987. This Table is based on data compiled by Rachel Schiller, Office of Basic Industries, International Trade Administration, US Department of Commerce.

Many of the agreements involve access to the products and processes of small US technology-driven firms which frequently lack the resources to produce and market their goods. In many ways, these smaller US companies also represent a mechanism for technology transfer at the international level. Seen from a US standpoint the question raised is whether ready access to US R&D by foreign companies is detrimental to US competitiveness now and in the long term (on the Japan-US relationship see, *inter alia*, Peters[53]). Seen from the opposite perspective of the countries whose MNEs are borrowing from the US science base, the question on the contrary is that of knowing whether these corporations will not start taking a further step with far reaching implications of gradually transferring a part of their corporate R&D base to the United States. The tendency by pharmaceutical MNEs to situate part of their research base in countries combining a good science-base and a large or fairly high sophisticated market was already noted for earlier periods[54]. In the 1960s and 1970s this process took the form of US investment in Europe, notably the UK. In the 1980s the trend has been strongly reversed with European companies investing in the United States and building up R&D capacities there rather than at home[55].

4. Present and currently predictable trade impacts in pharmaceuticals

In pharmaceuticals, biotechnology refers both to *products* and *processes*. It means *new processes* leading to improved old products, or to totally *new products*, but it also means improved traditional production *processes*, notably fermentation, along with their extension to the manufacturing of products which were previously obtained through organic chemical synthesis. Or again in human insulin it means a new genetic engineering route to a previous existing product obtained by extraction from animal sources.

The United States is practically the only OECD country where attempts have been made to assess sales and also the only one where such estimates are published by the government. The US Department of Commerce has reported on the difficulties it has met in making serious estimates. The *US Industrial Outlook 1988* states that "data on revenues (e.g. income or earnings) from products developed through biotechnology are neither sufficiently broken down in company reports nor collected by the government in enough detail to assess the impact of biotechnology on industrial sectors with precision".

Unofficial US estimates indicate that 1987 shipments of products developed through recombinant DNA and monoclonal antibody technology were around $550 million, an increase from $400 million in 1986, $220 million in 1985, and $60 million in 1984. Diagnostics utilizing monoclonal antibodies and DNA probes have been the fastest growing component of new biotechnology, accounting for more than half of the current market for new biotechnology-derived products. At least 220 diagnostic kits using monoclonal antibodies and 8 using DNA probes are now available. AIDS testing alone was valued at $75 to $100 million in 1987 and is growing rapidly. As a result of this rapid growth of sales, the marketing of new biotechnology products can now begin to be compared with current value of sales by the US pharmaceutical industry (*circa* $35 billion). $550 million already represents a little over 1.5 per cent of the pharmaceutical industry's present commercialised output.

In the case of pharmaceuticals the present low impacts on international trade patterns implied by the current value of output is reinforced by the particular pattern of production and commercialisation. The industry is one in which foreign direct investment has played an important part for many decades. Almost all companies of any significance have adopted a transnational form of organisation. They not only maintain sales forces in many countries, but manufacture drugs in a number of them and also carry out research abroad. Despite the low volume and high unit value of its products and their low transport costs, the pharmaceutical industry sends a surprisingly small proportion of its output across national frontiers. In 1980, world sales were about $85 billion in value, of which on $8 billion – less than 10 per cent – entered international trade in the form of finished drugs, and a further $6 billion in the form of intermediates. Local production by affiliates of multinational enterprises was over twice the size of direct imports in the year[56]. Despite the *potential trade creating effects* of new biotechnology, it is likely that *the growth of output of biotechnology* products and their commercialisation will, within the OECD area, probably *preceed* by many years *any significant impact on trade flows*.

Not surprisingly the International Trade Administration at the US Department of Commerce reports that while biotechnology has a small but *positive* impact on the US balance of payments, earnings came from a *wide range* of sources, which include royalties on licenses for products and processes, product sales, and R&D contracts funded by foreign firms or research organisations. An examination of the annual reports of some twenty biotechnology companies indicated that foreign earnings of one kind or another account for one-third to one-half of their income. In 1986-87 markets in Europe accounted for 80 per cent of these earnings, and Far East Asia accounted for 14 per cent[57].

5. Present and currently predictable trade impacts in agriculture

Although the unit value of the new biotechnology products is highest in pharmaceuticals, it is in agriculture that the single most important impact on trade has already been felt and that technological developments of considerable importance for trade are also in preparation. This is due to the fact that in agriculture the biotechnological tools liable to have an impact on production and trade draw on a wider range of techniques than genetic engineering and that given their origin in "modern" or "second generation" biotechnology, several of these technologies emerged in the early 1970s and have had time to develop identifiable impacts.

The new innovations have also emerged in conditions where, in the case of the developing countries at least, the economic and social groups affected have, in contrast for instance with the large pharmaceutical MNEs in the area of health care products, little or no capacity to control the *phasing* of the introduction of new processes and products. This is in part simply an expression of the present extremely inequal worldwide distribution of scientific and technological capacity in the area of biotechnology. But it also represents a new chapter in the long history of the relationship between agriculture and industry, within which the *initiative has always rested on the side of industry*. The latter has provided agriculture with vital inputs, but has also often been in a position to create industrial substitutes for the products supplied by agriculture. An earlier major chapter in this history was written by petrochemically based mass production of synthetic fibres, rubbers and plastics which developed partly on the basis of markets captured away from agriculture. *A new chapter has begun to be shaped on the basis of the techniques developed by biotechnology*. It already includes:

- *i)* The measurable impacts of advances in immobilized enzyme technology, through the extraction of fructose from starch;
- *ii)* Potential competition from single cell protein fermentation technologies;
- *iii)* The impacts arising from *in vitro* cloning and plant tissue culture on a growing number of crops used as raw material sources by industry.

In time, some of the impacts of these and comparable techniques may be mitigated by the development, through biotechnology, of new industrial uses of agricultural products. In addition, other technological advances due to biotechnology, e.g. enhancing nitrogen fixation in plants, may have a positive effect on developing countries' agricultural production and may also affect trading patterns.

a) *Enzyme-based extraction of fructose from starch and its impact on traditional sugar production and exports*

A technically feasible method for obtaining sweeteners from any starch, using enzymes as the biological agents (e.g. catalysts) in the production process, was developed at the beginning of this century. However, it was only with the advent of qualitatively improved enzyme-producing techniques and the availability of immobilized enzymes in the mid-1960s that the development of a maize-based sweetener known as high fructose corn syrup (HFCS) in the United States and as isoglucose in Europe became an economic proposition[58].

The driving force behind this technological development can probably be identified as the necessity by food-processing manufacturers, and notably those engaged in the bulk processing of basic raw materials to counteract the effects of the fall in profitability, by finding new diversified *industrial outlets* for their output of *products processed from agricultural raw materials.* Maize (or corn) was typically one such product and major US firms active in the processing of corn played a prominent role in the development of HFCS. High sugar prices in 1973 and 1974 and again in 1980 and 1981 provided a necessary market impetus leading to intense efforts to commercialize HFCS in the United States and, to a lesser extent, in Japan.

While HFCS is specifically based on maize, the technology permits the production of *starch derived sweeteners* from many raw material sources. Some Far East Asian countries may thus ultimately see the industrial processing of rice for the domestic sweetener market as an important outlet for their own surplus production once the "Green Revolution" has attained its full impact[59]. The substitution of HFCS for sugar stems from its technical properties, its nutritional (low calorie) advantages and its cost. The natural outlet for HFCS is in the industrial liquid sugar segment of the sweeteners market. In the United States, for example, in 1983, liquid sugar accounted for about 96 per cent of total industrial sugar use, or about two-thirds of total sugar use. Certain properties of HFCS permit better conservation of foods compared with sugar, and, being in liquid form, it also requires less labour for handling as compared with sold sugar in bags.

Production costs of HFCS have declined significantly over the past 10 to 12 years owing to energy savings and other improvements in manufacturing processes, and to a sharp reduction in the cost of enzymes. In mid-1985, HFCS sold at the dry-weight equivalent of 22 cents a pound in the United States, giving it a 30 per cent advantage over sugar both in the United States and Japan. A generally slower expansion of production of HFCS in the major sweetener consuming area of the EEC can be explained as much by economic as by political considerations. In contrast to the United States, the EEC has been (since 1976) a net exporter of sugar and depends heavily on production. Moreover, the profitability of HFCS as compared with refined beet sugar manufacture was sharply reduced after the mid-1970s by the removal of an existing subsidy to manufacturers of starch (the processed raw material input for HFCS) and the imposition of a quota-system on HFCS.

An exact assessment of the extent of the trade displacing impacts of HFCS on developing country sugar exporters would have to take into account other factors also determining the important shifts which have occurred in the sugar and sweetener trade: for instance the increased productivity of sugar beet production in Europe and the accelerated growth of synthetic chemical based sweeteners, notably the amino acid compound, *aspartame*. Some sound general indicators are available nonetheless:

i) In 1975, world consumption of HFCS was equivalent to only 500 000 tons of raw sugar, representing much less than 1 per cent of world sugar consumption. By 1985, HFCS had increased its share relative to world sugar consumption to more than 6 per cent. From 1980 to 1985, world production grew at an estimated 18.6 per cent per annum to nearly 6 million tons on a dry basis, of which 73 per cent was accounted for by the United States alone, followed by 11 per cent for Japan, 5 per cent for Western Europe and 3 per cent for Canada.

ii) In 1975, 90 per cent of internationally traded sugar came from developing countries. 70 per cent of traded sugar was imported by developed countries. By 1981, only 67 per cent of the internationally traded sugar came from developing countries, and only 57 per cent of the sugar that entered world trade was imported by developed countries: the United States and Japan alone are estimated to have cut back cane sugar imports by some 2.5 to 3 million tons.

iii) Another indicator of the trade displacing and/or trade reducing impact of HFCS and other industrial sweeteners is the trend of advanced countries' per capita sugar consumption (Table 8).

Table 8

APPARENT CONSUMPTION OF SUGAR IN TONS PER BILLION DOLLARS OF GDP OF OECD COUNTRIES

1961-63	1971-73	1981-83
6 154	5 141	3 682

Source: UNCTAD, 1986 (see reference 41).

In the future, growth of HFCS consumption and the process of substitution for sugar may be much slower than in the recent past. In the United States and Japan, based on current technology, the theoretical level of substitution of HFCS for sugar is close to the saturation level. In the United States, for example, the upper limit of substitution is estimated at 40 per cent of the total use of sugar and 60 per cent of industrial use. In other developed countries, the theoretical levels of HFCS substitution for sugar are lower than in the United States, because of a smaller industrial use of sugar, a smaller liquid sugar consumption within industrial sugar use and the political capacity of farmers' organisations to slow down the substitution process by obtaining appropriate legislation. However a rise in the market price of sugar which fell sharply in the early 1980s (from 62.3 US cents a kilo in 1980 to 12 cents in 1985), and further technological changes could alter the situation again. Experts now consider that with the development of crystallised fructose, further waves of substitution are likely, with inroads being made into domestic sugar consumption[60].

b) *Single-cell proteins from industrial substrates and potential competition with agricultural protein animal feeds*

Large R&D expenditures have been made by a number of major chemical and oil companies as well as by a few government laboratories for the large scale production of proteins, notably "single-cell proteins" (SCPs), through modern fermentation engineering technology[61]. Single-cell proteins are microbial proteins produced by the mass culture of yeasts or bacteria on hydrocarbons or other substrates (often waste by-products with few alternative uses). Their principal aim is to provide animal feed, while mycoprotein aims at providing human food. In order to make an impact outside special niches, single-cell proteins would have to compete with soya and fish meal. Exports to countries that need to reduce soya imports, or that combine a protein deficiency with cheap feedstock, have been considered to be attractive opportunities.

However, soya has remained such a cost-effective protein source that only very few companies in the OECD area (mainly in the UK which leads in this technology) maintain activities in single cell proteins, attempting to recoup some of their considerable investments. Other companies have left this field. The need to surmount regulatory hurdles and to gain consumer acceptance are substantial entry barriers. However, progress in this area may be possible by genetically engineering an organism to improve processing characteristics or by radically different means of fermentation, but the risk of technical failure is considered to be high. The situation with mycoprotein for human food, where small-scale consumer trials have taken place, is somewhat more encouraging.

If single cell proteins were competitive, their trade substituting and trade displacing effects could become considerable, in particular for major exporters of cattle meal.

It is still an open question as to whether the production of single-cell proteins would not make more economic sense for oil and gas producing countries as long as proteins could be grown on "free" natural gas. The conjectural entry of OPEC countries into the world protein markets would create a capacity to compete very seriously with countries that export cattle protein feeds (soybean meal in the case of the United States, Argentina and Brazil; peanut meal in that of Senegal and other West African countries). At present, the Soviet Union has built a capacity specifically aimed at decreasing its dependence on foreign sources of protein for animal feed. The USSR aims at producing a total of 3 million tons per year in order to be able to stop importing soybeans as a protein source.

c) *In vitro plant propagation and cell tissue culture*

In agriculture among the tools now offered by new biotechnology are the techniques of isolating cells, tissues or organs from plants and growing them under controlled conditions (*in vitro*). The advantages of *in vitro* propagation are:

i) Production of large numbers of plants or clones in a short time and using small confined facilities;
ii) Propagation of materials in an environment free of viruses and other pathogens, under optimum conditions and with better quality control and reliability.
iii) Ability to propagate plant species which are difficult to propagate vegetatively once they flower, or to propagate species where *in vitro* culture techniques are commercially superior to other conventional methods of propagation (e.g. for tropical plants with high level of tannin and phenols);
iv) Ability to supply plants on a year-round rather than on a seasonal basis, and from locations close to consumers and markets instead of being tied to cultivation sites;
v) Maintenance of heterozygozity and the cloning of superior individuals from both qualitative and quantitative viewpoints[62].

The techniques are still costly and of course require laboratory equipment, qualified scientists and agricultural engineers and special planting and propagation conditions. At a more general level relating to overall biological and environment phenomena, the effects of *in vitro* techniques tend strongly towards the *curtailment of genetic diversity* in trees and plants, a process the consequences of which cannot be fully measured today. However, from the standpoint of large firms engaged in food-processing or raw material extraction for chemical purposes, the immediate advantages offered by *in vitro* propagation are so strong that they overide longer-term considerations. Fruit trees, oil, date and coco-palm trees, species for lumber (including eucalyptus, aspen, poplar and pine trees) and pulp are good candidates for such *in vitro* cloning and have increasingly been the object of R&D. At present, flowers are the main crops propagated *in vitro*.

The example of palm trees

Outstanding results in *in vitro* cloning and subsequent planting of new plantations have been obtained with the palm species and allow for a discussion of potential trade impacts. The oil palm, which is grown exclusively in tropical developing countries, is important as both a cash crop (a quarter of the production is exported to OECD countries) and for food. Palm oil is extracted from the pulp and made into edible oils and margarine; cabbage palm oil is extracted from the stone of the fruit and is used in the manufacture of soap, detergents and cosmetics. Output doubled between 1960 and 1980, reaching 4.9 million tons in 1980-81, second only to the soybean (13.4 million tons) among oleagineous plants. World demand for oils has continued to increase rapidly, making further development of palm oil production a profitable aim.

The major problem in oil palm cultivation is that the trees rapidly grow too tall for practical harvesting, so that plantations have to be renewed every 25-30 years, requiring the propagation of millions of plantlets annually. Vegetative propagation is complicated by the fact that the oil palm tree bears male and female flowers which must be cross-pollinated, entailing a high variability in progeny and a lengthy selection process for successful crosses. As a result of the *in vitro* technique developed after considerable research, plantations with considerably higher productivity and easier harvesting conditions are coming to fruition.

The *outcome is renewed and increased competition between different raw materials* used in the oil and fat industry and hence between different branches of agriculture and the different producer and exporter countries, notably Third World countries situated in different continents. At present palm oil production in Malaysia and the Ivory Coast has been growing, mainly at the expense of coconuts, sun flower and cotonseed, and has already displaced a part of the exports of these products on world markets. Today processing in fats and oils is almost completely based on straight forward chemical engineering. It uses the continuous-production control and automation techniques developed in the chemical industry and enables practically any end-product to be made from raw materials. Firms engaged in this industry are thus posed to take full advantage from the competition waged by primary exporters. For them the progress in biotechnology represents in many ways simply a *new improvement* of an already dominant position *vis-à-vis* their supply base in agriculture[63].

Plant tissue culture aimed at modifying agricultural supply

In vitro cloning and propagation of trees is in reality only one area of a rapidly growing branch of the new agriculture-aimed biotechnology. Plant tissue culture offers increased possibilities of substituting agricultural specialities by industrially produced inputs. Many high-value plant-derived products used for pharmaceuticals, dyes, flavourings and fragrances are vulnerable to displacement as a result of current research. At present the production costs are sufficiently high so that plant tissue culture methods provide a viable alternative only for products valued above several hundred US dollars per kg[64], but increasing research and production experience will allow these costs to be reduced significantly. Data illustrating this are given in Table 9.

Table 9

AN ILLUSTRATIVE LIST OF PLANT TISSUE CULTURE RESEARCH ACTIVITIES[1]

Plant cultured	Plant products to be cultured	Plant country of origin	Research organisation	Value of product per kilogram ($US)	Market size ($US million)
Lithospermum	Shikonin	Korea, China	Mitsui Petrocheminal (Japan)	4 500	
Pyrethrum	Pyrethrines	Tanzania, Ecuador, India	University of Minnesota	300	20 (US)
Papaver	Codeine, opium	Turkey		850	50 (US)
Sapota	Chicle	Central America	Lotte (Japan)		
Catharantus	Vincristine		Canadian National Research Council	5 000	18-20 (US)
Jasminum	Jasmin	Many producers		5 000	0.5 (World)
Digitalis	Digitoxine-digoxine		University of Tubingen Boehringer-Mannheim (Germany)	3 000	20-55 (US)
Cinchona	Quinine	Indonesia	Plant Sciences Ltd. (United Kingdom)		
Cacoa	Cocoa butter	Brazil, Ghana	Cornell University Hershey, Nestlé		891 (World)
Thaumatococcus	Thaumatine	Liberia, Ghana, Malaysia	Tate and Lyle United Kingdom		
Rauwolfia	Reserpine				80 (US)

1. These data are presented for illustrative purposes only. While data are accurate to the best of our knowledge, it should be kept in mind that the proprietary nature of most of this research makes it difficult to achieve access to the latest data. All but two of the plant tissue culture activities listed above are at the research stage; Shikonin has already been commercialised, and Digitalis is currently at the scale-up stage.

Source: Kenney and Buttel, F., "Biotechnology : Prospects and Dilemmas for Third World Development", *Development and Change*, Vol. 16, No. 1, 1985.

The impact of successful production of substitutes will be felt by countries dependent upon exports of the natural products concerned, in which often they have hitherto enjoyed resource endowment comparative advantage because of the specific conditions required for growth. As noted earlier such displacement has already occurred in the past as synthetics have been developed (e.g. the Indian indigo industry, jute in Bangladesh and steroids in Mexico). Plant tissue culture must however be understood as representing a far greater source of competition and substitution because the techniques can be applied to any plant, whereas there were great technical difficulties in synthesizing many plant products by conventional synthetic techniques.

VII. PROSPECTIVE IMPACTS ON EMPLOYMENT

1. Introduction

a) *Recent experience with technology and employment*

A large and growing literature discusses the impacts of technology on employment, particularly the question of whether and how new technologies have affected jobs in industrialised countries since the Second World War. In the late 1970s, the beginning of the "microelectronics revolution" coupled with the economic downturn fuelled concern that widespread use of microelectronics would have a labour-saving bias, and thus lead to a period of "jobless growth".

Numerous studies indicate that global employment levels in the OECD area in recent years have not been significantly influenced by technological progress, and that macro-economic factors, particularly growth rates, shifts in demand patterns and international competition have been much more important[65]. New technology has brought about both job-gains and job-losses, with gains apparently exceeding the losses. Employment impacts have varied considerably between economic sectors. Technological progress has contributed to job losses in the manufacturing sector, although employment in high-technology industries has increased in a few countries without, however, changing the negative net balance. In the service sector, technological progress has been accompanied by the creation of new jobs, particularly in business, financial and communication services. The sum of these new jobs more than compensates for the technology-induced losses incurred by the manufacturing sector. The major effect of technological progress has been felt less in total employment levels, than in changes in the structure of employment and in higher skill requirements.

So far, studies have focused almost exclusively on the impacts of information technologies, with biotechnology scarcely mentioned.

Industry and governments have not linked their support for biotechnology to hopes that its development could create many new jobs. In fact, the few surveys conducted in the 1970s shed doubts on the labour-creating potential of the new biotechnologies at least during the first years of development[66]. Persistent manpower shortages have, since 1987, led to a somewhat more optimistic assessment of this labour-creating potential, both in the short and long term, at least for well-trained people.

Nevertheless, trade unions and agricultural interest groups have expressed concern that biotechnology could have negative employment effects.

b) *Current employment related to biotechnology*

In view of the economic size of the sectors which will be affected by biotechnology, particularly their importance for jobs, the employment question is a legitimate one. Even excluding the possible impacts on sectors which biotechnology might penetrate at a later stage (such as energy, or mineral extraction) the number of jobs in the sectors which could be affected is large. Tables 10 and 11 give employment in the OECD area, as a percentage of total employment, in the sectors which will be first and most affected by the new biotechnologies: agriculture, public health, and the food and chemicals industries (including pharmaceuticals). Large international variations can be found reflecting mainly the variations in agricultural employment between OECD countries. However, even in the most industrialized countries, employment in all sectors together exceeds 10 per cent of total employment, in at least 9 countries (Finland, France, Iceland, Italy, Japan, New Zealand, Norway, Spain, Switzerland) it is approximatively a quarter or a fifth of all employment, and in three countries (Greece, Portugal, Turkey), it is much higher due to the large size of their population still active in agriculture.

It is more difficult to find data on employment in biotechnological activities or industry directly. Estimates in the 1970s indicate that in the United Kingdom and the Netherlands, between 20 and 25 per cent of the production of the food and beverage industries comprises fermented products. If it is assumed that manpower percentages in this sector do not deviate

Table 10

EMPLOYMENT IN SECTORS WHICH WILL BE AFFECTED BY NEW BIOTECHNOLOGY

As percentage of civilian employment; 1983 unless dated otherwise

Countries	Agriculture	Health	Chemical industries	Food industries	Total
Australia	6.6	7.0	–	–	–
Austria	8.6	5.4 (1982)	2.0	2.1	–
Belgium	3.0	4.6 (1981)	2.7	3.1	–
Canada	5.5	4.9 (1981)	–	–	–
Denmark	7.4	4.8 (1980)	1.5	3.1	–
Finland	12.7	5.8	–	–	–
France	7.9	6.2	2.6	2.9	19.6
Germany	5.6	2.5	3.5	1.8	13.4
Greece	29.9	2.0 (1981)	–	–	–
Iceland	10.7	6.9 (1979)	1.0	12.3	–
Ireland	17.1	5.2	–	–	–
Italy	12.4	3.0	1.8	1.1	18.3
Japan	9.3	3.1 (1981)	3.7 (1979)	2.2 (1980)	–
Luxembourg	4.7	3.8 (1981)	–	–	–
Netherlands	5.0	6.5	2.4	2.7	16.6
New Zealand	11.2	6.0	1.9	5.9	25.0
Norway	7.5	6.7	1.4	2.8	18.4
Portugal	23.6	2.4	1.5 (1980)	2.2 (1980)	–
Spain	18.6	3.3	2.1	3.2	27.2
Sweden	5.4	7.9	1.6	1.6	16.5
Switzerland	6.7	5.5	2.9 (1980)	3.8 (1980)	–
Turkey	58.9	–	–	–	–
United Kingdom	2.7	5.3	2.9 (1978)	3.2 (1978)	–
United States	3.5	5.8	1.7	1.5	12.5
Total OCDE	9.5	3.6[1]	1.3[1]	1.2[1]	–

1. Refers only to countries for which 1983 data are available.

Sources: *Labour Force Statistics 1964-83*, OECD, Paris, 1986. *Historical Statistics 1960-85*, OECD, Paris, 1987 (Agriculture). *Measuring Health Care*, OECD, Paris, 1985 (Health). OECD Database Sector ISIS, in segment ISIC (Industry).

Table 11

EMPLOYMENT IN SECTORS WHICH WILL BE AFFECTED BY NEW BIOTECHNOLOGY

As percentage of civilian employment, in countries where comparable figures for all sectors are available for other years than 1983)

Countries		Agriculture	Health	Chemical industry	Food industries	Total
Austria	(1982)	8.7	5.4	2.0	2.1	16.2
Belgium	(1981)	3.0	4.6	2.7	3.1	13.4
Denmark	(1980)	7.1	4.8	1.5	3.0	16.4
Iceland	(1979)	11.7	6.9	1.0	3.3	23.1
Japan	(1978)	11.7	2.8	2.0	2.3	18.8
Portugal	(1980)	27.3	2.2	1.5	2.2	33.2
Switzerland	(1980)	6.9	5.1	2.9	3.8	18.7

Sources: See Table 10.

widely from production percentages, then relatively large numbers of people are employed in classical biotechnology manufacturing. In the pharmaceutical industries, an estimated 25 per cent of all products go through a fermentation process, including antibiotics and some vitamins (mostly modern biotechnology), but it would be hazardous to take this as a basis for a manpower breakdown in the drug sector. The numbers employed in genetic engineering companies in the United States are better known; trade associations published figures of more than 30 000 employees working in these companies in 1982-83[67], and of approximately 40 000 in 1987. This figure is perhaps an underestimate as it probably does not include the manpower of some large corporations working on new biotechnology projects.

2. Categories of potential effects

In the absence of comprehensive data on present employment in biotechnology companies, numerical estimates of future employment effects in the biotechnology industry, not to speak of the indirect effects, are very hazardous. However, the employment strategies of biotechnology industries in the recent past are known to some extent, as are many technical and economic trends in the sector. Comparing those to the employment effects of other technologies which have been better studied, it is possible to draw up a number of possible employment trends due to biotechnology developments.

To this end, it is useful to break down *quantitative* employment effects into the following *four* categories:

– *Direct effects*

a) Suppliers (biotechnology companies);
b) Users (agriculture, health, industry);

– *Indirect effects*

c) Investment multipliers (employment through new capital investment);
d) Costs, income, demand (employment through higher demand resulting from cost-reductions).

a) *Direct effects: suppliers*

The situation in 1968-87, emerging from the interviews with industrial companies (Chapter II, 4*a*) has been characterised by large R&D expenditures, and a small number of new products. As a consequence, the evaluation of possible returns on investment by biotechnology companies has become more conservative and has led, in a large majority of all visited companies (70 in a total of 94), to an intensive search for rationalisation, that is for cost-reducing measures, and these include labour costs. Most companies have made it clear that they have not introduced biotechnology in order to create jobs, but to get cheaper products through process-rationalisation, and to improve their market position. The rationalisation drive shows itself in a policy to optimise the available manpower, that is in a policy of in-house retraining.

If one focuses on ***R&D manpower*** alone, a more encouraging picture is emerging. It is revealing that, in more than one country, the R&D personnel of biotechnology companies which have disappeared or which had to cut jobs, have easily and quickly found new employment. The "science-push" of the last years has produced an expansion of industrial R&D manpower, and even if this has led to temporary overcapacities in a few cases, new shortages of skilled manpower have lately appeared in the biotechnology industry. Thus, the R&D manpower expansion is expected to continue into the 1990s.

However, the large costs of in-house R&D manpower, due to their high qualification profile, have also encouraged companies to look for cheaper and more flexible and cost-efficient ways to build up knowledge. Such ways are the funding of external research, particularly at universities, and the forging of links with small R&D companies for often limited periods. Together, industrial biotechnology activities in the OECD area have created or supported tens of thousands of jobs in all parts of the research system.

In the *future*, various employment trends will make themselves felt in the bio-industries. They will, to no small degree, be a function of innovation trends and technical change in the bio-industries. There are particularly four aspects of the future direction of innovation in biotechnology which will have consequences for employment:

i) *Process versus product innovation*

Process innovation has often been found to be job-saving, and product innovation job-creating. In the short term, biotechnological innovation might focus more on new processes than on new products, although this must not in all cases lead to job losses. In the longer term, the balance between new processes and new products could shift in favour of the latter. It is, in this respect, encouraging that 65 per cent of the interviewed companies with projects or plans in biotechnology have replied that they intend to develop new products; and 25 per cent new processes (Chapter II, 2*a*);

ii) *Substitution of old versus creation of new products*

Substitution of old products has often been found to be job-saving. The first biotechnology products with applications in medical therapy or agriculture have mostly been substitution products (human insulin, interferon, artificial sweeteners and others). In the short term, there will be substitution effects of new biotechnology innovations in industry and agriculture, including effects on employment. In the longer term, one must keep in mind the growing potential of biotechnology to generate unexpected new products; monoclonal antibodies, gene-probes, and genetic fingerprinting already testify to this potential which might shift the balance from substitution towards new products;

iii) *Labour versus capital bias in innovation*

Most technologies have a factor bias; usually they tend to substitute labour by capital. In other terms, there is a general tendency towards automation and standardisation which is labour saving. It is true that this tendency can be felt in the bio-industries as well, but there are two technical factors which might, in the longer term, limit its negative employment effects.

First, there are some limits to automation due to the nature of important new biotechnology products and processes. Biological components, particularly proteins, have a much more complicated molecular structure than the products hitherto manufactured by the pharmaceutical industry. Thus, they call for complex production, purification and control techniques which do not always allow for labour-saving standardisation. Similar effects can be found in cell and tissue cultures which call for precise controls carried out by highly skilled people. Also, the development of tailormade plants with increasingly complex qualities could be relatively labour-intensive. Second, even if biotechnologies are found to be labour-saving, they must not necessarily be capital-intensive. Their factor-saving potential can extend to other factors as well, with considerable macroeconomic consequences (Chapter VII, 2*d*).

iv) *Small company versus big company innovation*

Empirical evidence suggests that small establishments create more employment than big ones. In the United States, 80 per cent of all jobs created during the last years are due to companies with less than 100 employees. The factors which currently favour the exploitation of biotechnological innovation by big rather than small companies (Chapter II), will reinforce concentration and labour savings in the biotechnology industry. In the longer term, with the possible entry of smaller companies into the sector ("bandwagon effect"), employment prospects could improve again.

Apart from these aspects of innovation, shifts in output and demand patterns will affect employment in the biotechnology supplier companies. Such shifts could have output and labour reducing consequences particularly in pharmaceutical and food industries. The often mentioned move (Chapters II, V) towards early diagnosis and prevention, perhaps accompanied by one towards more self-medication, could bring about a reduction of output and employment in pharmaceutical industries where therapeutic drugs constitute the bulk of production. Delays in the introduction of the new products, growing markets for diagnostics and vaccines (prevention) and the emergence of new, or the return of known diseases will partly compensate for this reduction. But ultimately, the difference between the costs of producing a one-time vaccine to prevent a disease, and the costs of producing the drugs to treat this disease sometimes over years, can be so huge, that even a moderate shift towards early diagnosis and prevention might have negative employment consequences in the drug industry.

What has been discussed until now, has been *quantitative* employment impacts. According to many experts, however, it is the *qualitative* impacts which, in time, will be most important, and their direction is already more clearly visible than that of the quantitative impacts.

Table 12

EMPLOYMENT IN AGRICULTURE

As percentage of civilian employment 1978 and 1985

Countries	1978	1985
Australia	6.41	6.17
Austria	9.63	8.15
Belgium	3.19	2.91
Canada	5.75	5.22
Denmark	7.85	6.70
Finland	14.43	11.54
France	9.19	7.56
Germany	6.10	5.44
Greece	32.02	23.90
Iceland	12.87	10.34
Ireland	20.64	15.86
Italy	15.45	11.20
Japan	11.70	8.77
Luxembourg	6.41	4.38
Netherlands	5.38	4.93
New Zealand	11.24	11.14
Norway	8.68	7.30
Portugal	31.26	23.15
Spain	20.64	18.24
Sweden	6.10	4.45
Switzerland	7.30	6.62
Turkey	60.69	57.06
United Kingdom	2.75	2.55
United States	3.70	3.12
Total OCDE	10.62	8.88

Source: See Table 10.

Qualitative impacts will result from the responses to current training and manpower needs in biotechnology which continue to be a source of concern for governments and industry alike and which have, therefore, been scrutinised in various countries[68]. There is agreement that a high qualification profile is a predominant feature of the manpower needs in biotechnology and that, so far, industry, the university and governments have not been really successful in addressing these needs. Improved training in various disciplines and, more importantly, *retraining* of the existing labour force will raise the skill profile not only in the biotechnology supplier companies, but in the user sectors as well; amongst numerous examples are the need to retrain doctors and patients in the use of doctors' and home tests, and to develop extension services in agriculture to assist farmers in the application of biotechnologies.

b) *Direct effects: users*

The diffusion of biotechnology through the economy, and the productivity increases this will bring about, will have larger employment effects than those which can be expected in the bio-industries themselves. New biotechnology products and markets will lead to demand widening which could be felt across the economy.

Table 10 has shown that agriculture, the health sector, and the chemical and food industries will be first and foremost affected. These are also sectors where resistance to change is often high, one more reason why no numerical forecasts of general employment effects of biotechnology are possible.

Of the numerous possible employment effects of biotechnology, those which can be expected in *agriculture* are, without doubt, the most critical ones, both politically and economically. They are the only *prospective* employment effects of biotechnology which have already provoked effective countervailing forces. Agriculture has, in the last years, adopted biotechnology more slowly than would have been technically possible because a perceived threat to employment has acted as a brake on diffusion. Still notorious is the 1979 decision of the EEC to impose a quota system on isoglucose, a biotechnology-derived sugar substitute, to protect the European Community's more than 300 000 sugar beat farmers. This decision has seriously curbed the diffusion in Europe of one of the first promising and competitive food products based on modern biotechnology. The European consumer still pays approximately twice the 1987 world market price for his sugar and this price has in part been maintained by lack of availability of the cheaper sugar substitute.

Present difficulties with bovine growth hormone in Europe have various explanations, one of them being the agricultural labour-saving potential of a product which could increase milk production considerably.

How justified are the fears of job reducing effects of biotechnology in the agriculture of OECD countries? Taking a historical perspective, one can argue that many OECD countries have their biggest agricultural adjustments long behind them. Table 13 shows, in the case of six countries, that these adjustments, characterised by a dramatic growth of output per person, and by large annual job losses, have mainly taken place in the 1950s and 1960s, and that already in the 1970s, annual job losses tended to become smaller in some countries. Table 12, which compares agricultural employment as a percentage of civilian employment between 1978 and 1985, tends to confirm this general trend. In some countries with still large agricultural populations in the 1970s (i.e. Finland, Greece, Ireland, Italy, Portugal), very substantial reductions of agricultural employment (up to a third in relative terms) have taken place in the seven years between 1978 and 1985, before new biotechnology could have had much of an impact.

Agricultural productivity will keep rising, employment will decline further and agricultural adjustment will, thus, remain a continuous, long-term process, with or without biotechnology. What then is the additional pressure which new biotechnology might exercise on this process, and particularly on jobs? Answers are likely to vary considerably between sectors and countries; the Mediterranean and other OECD countries with large agricultural sectors will feel more pressure than the industrialised countries. Obviously, biotechnology could facilitate agricultural adjustment if it were to concentrate its efforts more on quality improvements and the development of new, industrially useful crops (Chapter V, 3*b*) rather than on further agricultural production increases. However, even if biotechnology does increase quantities as well, it will do so in the context of other, parallel technological advances which act together.

G. Junne's examination of the possible impacts of bovine growth hormones in Europe is an interesting case study because the figures are perhaps more widely significant for the relative weight of biotechnology impacts, particularly in the employment context[69]. The dairy sector is the most important subsector of European agriculture, accounting in 1984 for almost 19 per cent of the total value of agricultural output in the EEC area. It has been calculated

Table 13

GROWTH OF OUTPUT, PRODUCTIVITY AND EMPLOYMENT IN AGRICULTURE – 1950-78

Annual average compound growth rates

GROWTH OF OUTPUT

	1950-73	1973-78
France	2.0	0.1
Germany	2.3	1.0
Japan	3.2	–1.0
Netherlands	3.1	3.2
United Kingdom	2.6	0.9
United States	1.9	0.9
Average	2.5	0.9

GROWTH OF OUTPUT PER PERSON EMPLOYED

	1950-73	1973-78
France	5.6	5.4
Germany	6.3	5.0
Japan	7.3	–1.2
Netherlands	5.5	4.9
United Kingdom	4.7	2.8
United States	5.5	1.2
Average	5.8	3.4

GROWTH OF EMPLOYMENT

	1950-73	1973-78
France	–3.5	–4.2
Germany	–3.7	–3.8
Japan	–3.8	–2.1
Netherlands	–2.3	–1.7
United Kingdom	–2.0	–1.9
United States	–3.5	–0.3
Average	–3.1	–2.3

Source: Angus Maddison, *Phases of Capitalist Development*, Oxford, New York, 1982, p. 117.

that the combined effect of continuous upgrading of breeding, improved feed conversion, progress in veterinary sciences and use of growth and other hormones, will add up to tremendous productivity increases. If total production volume is not allowed to increase, approximately 33 per cent of the current livestock would have to be taken out of production by the year 2000. Even if bovine growth hormones were completely banned in Europe, the necessary reduction in cattle numbers will still be 22 per cent, and many cattle farmers will have to leave the sector. Thus, the prohibition, even of one of the most important new agricultural biotechnology products could delay, but not stop an apparently inevitable evolution, because of the ongoing, synergistic effect of other innovations on productivity.

The first obstacles to the penetration of agriculture by biotechnology innovations appeared almost ten years' ago. In the meantime, agricultural opportunities and consequences of biotechnology have become better known and new trends have emerged which may point to political and institutional changes:

- There is growing political awareness in OECD countries of the high economic costs of agricultural protectionism and increasing pressures to have agriculture more directly exposed to the forces of competition and technical progress. In the new agricultural policy context which is appearing on the horizon of the 1990s, biotechnology could become a major instrument of change as has been indicated in Chapter V;
- There is a growing biotechnology awareness in ministries of agriculture and agricultural research centres, coupled with growing readiness to accept the new cost-reducing innovations.

In order to compensate for job losses and enhance employment prospects in agriculture, the use of biotechnology for the development of new, industrially useful high-value crops has been proposed (Chapter V, 3*b*). Although it is doubtful that many economically viable projects have been found so far, such ideas indicate a growing convergence of industrial and agricultural policies and practices, which by itself, would constitute one of the social and institutional innovations which are a precondition for biotechnology diffusion.

Agricultural employment effects of biotechnology will also depend upon a country's international trade in agricultural products. A leadership position in agricultural biotechnology innovation, translated into increased exports, may create jobs. Imports of biotechnology products may lead to loss of jobs.

How will biotechnology affect employment prospects in the *public health* sector? In the short run, the development of an increasing number of cheap and fast diagnostic tests may reduce employment in bio-medical laboratories. In the longer term, it is doubtful whether this or any advance in disease prevention and cure will have a really major impact on health employment. The latter will be more influenced by the general ageing of the population which will maintain increasing pressure on the health system. Demands for better health care and growing public awareness of opportunities arising from research, will accelerate diffusion of innovation, irrespective of employment impacts. In any event, even if biotechnology would reduce jobs in the health sector, which is far from certain, this would not be viewed primarily as a threat to employment but rather as a most welcome relief for public health budgets, contrary to the attitudes prevailing with regard to agricultural employment.

c) *Indirect effects: investment multipliers*

Development, production and installation of equipment for the manufacturing of new products result in one of the main indirect job-creating effects of new technology. As industry has invested during the last years in biotechnology developments across the OECD area, such indirect employment effects are already occuring. They will continue as investments are planned to increase in the coming years. Few figures have been given for biotechnology investments. One source mentioned $400 million as global figure for investment in research and production *equipment* in 1986[72].

d) *Indirect effects: costs, income, demand*

Technical progress leads to productivity increases, reduced production costs and hence to higher profits or wages, or to lower prices. This will increase real income and demand which is likely to be translated into higher employment in the economy in general. Thus, employment reductions in one sector, if they are due to general factor saving effects and not compensated for by increased capital costs, will after a time-lag, theoretically result in higher employment in other sectors. In conditions of competition, productivity increases and factor-saving effects will find their way into price reductions. In monopolistic and oligopolistic conditions, productivity increases will find expression in higher profits and/or wages.

This report has mentioned the production costs of biotechnology, several times, as they have important implications for a number of economic issues. One is economic pervasiveness. Chapter III, 4*b* has shown that cost reductions (for new processes rather than new products) are one of the conditions of wide diffusion of a new technology, and that the costs of new biotechnological processes in key sectors such as bulk chemicals, are still too high for successful competition with traditional processes. Second, costs have consequences for the structure of industry. Chapter II has pointed to a concentration trend in the biotechnology industry and to widespread co-operation between large and small companies, which is explained by currently high research and marketing costs, and by a low return on investment due to the still small number of profitable products.

This chapter looks at costs from a third economic perspective, that of employment consequences. In order to understand the possible long-term impacts of biotechnology, it is not enough to assess the short-term employment effects in isolation from other, perhaps longer-term factor-saving effects. The question is whether present labour-saving tendencies will be matched by equivalent capital and other factor-saving tendencies, that is whether biotechnology will be less costly in terms not only of labour, but also of capital, materials, land or energy.

No comprehensive reply to this can be expected from industry, because experience is still limited, large differences between sectors, and investments in biotechnology not always separated from other investments (Chapter II, 2*c*). Also, there is concern in industry that premature statements about possible cost reductions might create unrealistic expectations, and that such reductions could be offset by high safety and other overhead expenditures.

The published literature reflects the same difficulties. Discussing the economics of biotechnology, A. Hacking refers both to the current uncertainties in assessing costs, particularly of fermentation, and to the potential for further cost-reducing improvements[70].

However, there are already more than a few examples of factor-saving effects of biotechnology which go beyond labour. Many result from the "dematerialisation" trend (Chapter V, 1*b*) whereby traditional raw materials are replaced by rDNA derived products and processes.

The, perhaps, extreme example of thymus-hormone was given in 1985[71]. Thymus-hormone is produced from calf thymus glands by extraction, ultrafiltration and chromatography. The global supply of calf thymus glands is limited to approximately 50 tons per year, which is much less than would be necessary for worldwide therapeutic treatment. The production costs for 12 mg which is the necessary therapeutic dosage per person, amount to $1 000 (1985). With the help of rDNA technology, the same product could be manufactured without any raw material supply limitations and calculations indicate that production costs could come down from $1 000 to less than $1 for one therapeutic dosage, during the 1990s.

Other examples have emerged since, indicating capital-cost reductions in the rDNA production of insulin, human growth hormone, or interferons. The expression of interferon in genetically modified micro-organisms, for example, is today much more efficient than had been anticipated only a few years ago.

In plant and agricultural biotechnology, new cell culture techniques have the potential to save considerable amounts of land and water, as they allow for the growing of plants in laboratory and hothouse conditions, and sometimes without any land. In some countries, the land-saving component of modern biotechnology might be as important as any other factor savings.

In time, the capital and other factor-saving effects of biotechnology could in some sectors be very substantial and exceed labour-saving effects. In any event, the beneficial macro-economic effects of biotechnology through general factor savings, will take time to be felt, and they may not always emerge in corresponding price reductions.

If the term *indirect* is used in a broader sense, other indirect employment effects of biotechnology, which have not been discussed here, could be envisaged. Most of these derive from the disease-reducing health impacts of biotechnology which will improve the working capacity and prolong the potential working life of the population, thus increasing the numbers of employed, or of employment-seeking people.

3. Conclusions

The net result of so many possible developments with both positive and negative employment effects, cannot be precisely evaluated.

During the next ten to twenty years, new biotechnologies will develop a considerable factor-saving potential. There will be opportunities for labour-savings which in some instances will be particularly attractive, due to the substitution character of new innovations, especially in agriculture, and the cost-reduction policies of the large industrial companies active in biotechnology.

Countervailing forces, particularly in agriculture, might again delay the diffusion of biotechnology by a few years, precisely because of these potentially labour-saving effects, as they have done in the recent past. Otherwise, biotechnology might during this period, add to the unemployment problem. The figures could vary considerably between countries, depending upon the size of their agricultural sectors.

The countries least vulnerable to negative job-impacts of biotechnology are those where technical progress has already had large-scale effects and where employment in agriculture is comparatively small. They all belong to the group of countries which also have the most advanced biotechnology industries: Germany, the Netherlands, the UK, the US, Sweden, Switzerland, among others.

In these and other industrialised countries, the additional job-losses due to new biotechnology alone, if any, will be quite small in relation to total unemployment. In many cases, these losses might be compensated for by new employment created by new biotechnology products and markets, or by expanding biotechnology exports. In other countries, including the Mediterranean OECD countries, job losses could be somewhat larger. In some Third World countries which depend upon monocultures that have become replaceable through biotechnology (e.g. sugar), employment consequences might be heavy and difficult to absorb (Chapter VI). However, it must be stressed here that there are many other technological factors at work which will inevitably reduce agricultural employment in these as well as in OECD countries (fertilisers, mechanisation) and that, in general, it will not be possible to evaluate the impacts of biotechnology in separation from other factors. In any event, trade and competitiveness will substantially influence the employment effects of biotechnology in each country. While restructuring may be a prerequisite for industrial competitiveness, the most important, long-term impacts on employment will reflect competitive positions and are therefore likely to *follow* changes in trade patterns.

During the next twenty years, other factor-saving effects of biotechnology, particularly capital-, land- and materials-savings, will become increasingly noticeable, in certain cases proportionate to, or even greater than labour-savings.

Biotechnology is likely to continue its rapid scientific and technical progress during this period. This progress will help create the conditions which could, in more than 10 years from now, turn biotechnology into a net creator of new jobs. In industry, the number of new products, rather than substitution products is expected to increase; in agriculture, biotechnology might contribute to the development of new, useful crops to replace those which have been made redundant; in environment protection, an increasing number of new biotechnology applications could help create employment.

What is certain is that the skills in the industries and sectors connected with biotechnology will undergo profound changes in the direction of a higher qualification profile, providing a growing number of jobs to well qualified people. Many of these will have acquired their qualification through retraining.

To judge from past experience, more than ten years will be necessary for the social, institutional, legal and government policy adjustments without which biotechnology diffusion will be slower than technically and economically possible. Some of these adjustments have begun already, and the best strategy to help overcome negative employment effects in the long run is to accelerate, rather than to delay them.

After the turn of this century, many of the most pressing adjustments might be completed. By then, the technology itself may be ready for further fundamental advances, including in the bio-sensor, bio-chip and neuro-computer areas, in large-scale diagnosis and prevention of diseases, in the cure of genetic and other grave diseases, and others.

At this point, biotechnology may begin to play an economic and social role more comparable to that of information technologies; it might become a dynamic, job creating factor in the increasingly service-based economies of the OECD countries.

NOTES AND REFERENCES

1. *Social Dimensions of Biotechnology – Toward a European Policy*, Papers presented at an International Seminar in Dublin, 12th-13th November 1987.
2. Théodore Friedman, M.D.: *Gene Therapy, Fact and Fiction in Biology's New Approaches do Disease*, a Banbry Public Information Report, Cold Spring Harbor Laboratory, New York, 1983.

   Human Gene Therapy, Background Paper, Congress of the United States, Office of Technology Assessment, Washington, D.C., December 1984.

   F. Anderson, "Le Traitement des maladies génétiques", in *La Recherche*, No. 176, Paris, April 1986.
3. *Technical Change and Economic Policy*, OECD, Paris, 1980.
4. *The OECD List of Social Indicators*, OECD, Paris, 1982; *Living Conditions in OECD Countries – A Compendium of Social Indicators*, OECD, Paris, 1986.
5. *Transnational Corporations in Biotechnology*, United Nations Center on Transnational Corporations, New York 1988, pp. 14-18.
6. Andrew J. Hacking, *Economic Aspects of Biotechnology*, Cambridge University Press, 1986, pp. 256-257.
7. *Commercial Biotechnology – An International Analysis*, Congress of the United States, Office of Technology Assessment, (OTA), Washington D.C., 1984, p. 140.
8. National Science Foundation, SRS 18th March, 1988: NSF 88–306; *New Developments in Biotechnology: US Investment in Biotechnology – Special Report*, OTA-BA-360, Washington D.C., US Congress, Office of Technology Assessment, July 1988, pp. 77-94.
9. On these and related issues, see also: Freeman, C., "The Challenge of New Technologies", OECD 25th Anniversary Symposium on *Opportunities and Risks for the World Economy: The Challenge of Increasing Complexity*, OECD, Paris, 1986.

   Freeman, C., *Technology Policy and Economic Performance: Lessons from Japan*, Frances Pinter, London, 1987.

   Freeman, C. and Perez, C., "Innovazione, Diffusione, e Nuovi Modelli Tecno-Economici", *L'Impresa*, No. 2, Milano, pp. 7-14, 1986.
10. Katz, B.G. and Phillips, A., "Goverment, Technological Opportunities and the Emergence of the Computer Industry", in Giersch, H. (ed.), *Emerging Technologies consequences for Economic Growth, Structural Change and Employment*, J.C.B. Mohr, Tübingen, 1982.
11. Diebold, J., *Automation: The Advent of the Automatic Factory*, Van Nostrand, New York, 1952
12. Devine, W., "From Shafts to Wires: Historical Perspectives", *Journal of Economic History*, Vol. 43, pp. 347-373, 1983.

13. Mansfield, E., "Technical Change and the Rate of Imitation", *Econometrica*, Vol. 29, No. 4, pp. 741-766, 1961.
14. Metclafe J.S., "The Diffusion of Innovation in the Lancashire Textile Industry", *Manchester School*, No. 2, pp. 145-162, 1970.
15. Schumpeter, J., *Business Cycles*, 2 volumes, McGraw Hill, New York, 1939.
16. Nelson, R.R. and Winter, S.G., "In Search of a Useful Theory of Innovation", *Research Policy*, Vol. 6, No. 1, pp. 36-76, 1987.
17. Dosi, G., "Technological Paradigms and Technological Trajectories", *Research Policy*, Vol. 11, pp. 147-163, 1982.
18. Kuhn, T.S., *The Structure of Scientific Revolutions*, Chicago University Press, 1982.
19. Perez, C., "Structural Change and the Assimilation of New Technologies in the Economic and Social System", *Futures*, Vol. 15, No. 4, pp. 357-375, 1983.

    Perez, C., "Microelectronics, Long Waves and World Structural Change: New Perspectives for Developing Countries", *World Development*, Vol. 13, No. 3, pp. 441-463, 1985.

    Perez, C., See Freeman and Perez, 1986.
20. Rehm, H.J., "The Economic Potential of Biotechnologies", Kiel Conference on Emerging Technology, Kiel Institut für Weltwirtschaft, published in Giersch, H. (ed) (1982), *Emerging Technologies: Consequences for Economic Growth, Structural Change and Employment*, J.C.B. Mohr, Tübingen, 1982.
21. Cooney, C.L., "Biochemical Engineering Solutions to Biotechnological Problems" in *New Frontiers of Biotechnology*, NAC, Washington, D.C., 1984.
22. Postgate, J., "Microbes, Microbiology and the Future of Man", *The Microbe*, pp. 319-333, Society for General Microbiology, Cambridge University Press, 1984.
23. Warhurst, A., "The Potential of Biotechnology for Mining in Developing Countries: the Case of the Andean Pact Copper Project", D.Phil. thesis, University of Sussex, 1986.
24. Faulkner, W., "Linkage between Academic and Industrial Research: the Case of Biotechnological Research in the Pharmaceutical Industry", D. Phil. thesis, University of Sussex, 1986.
25. Kristensen, R., *Biotechnology and the Future Economic Development*, Institute for Future Studies, Copenhagen, 1986.

    Hollander, S.G., *The Sources of Increased Efficiency: A Study of Du Pont Rayon Plants*, MIT Press, 1965.
26. *Biotechnology and the Changing Role of Government*, OECD, Paris 1988; and *Recombinant DNA Safety Considerations*, OECD, Paris 1986.
27. Karl Heusler, "The Commercialisation of Government and University Research, and Public Acceptance of Biotechnology", Canada-OECD Joint Workshop on National Policies and Priorities, in *Biotechnology and the Changing Role of Government*, *op cit.*, p. 109.
28. For a summary, *Assessing the Impacts of Technology on Society*, OECD, Paris, 1983.
29. Mark F. Cantley, "Democracy and Biotechnology", in *Swiss Biotech*, May 1987, No. 5, pp.5-15.
30. Jeremy Rifkin.
31. The latest are in *Public Perceptions of Biotechnology*, US Congress, Office of Technology Assessment, Washington, D.C., May 1987.
32. *Biotechnology and Patent Protection – An International Review*, by F.K. Beier, R.S. Crespi, J. Straus, OECD, Paris, 1985.
33. Betty Dodet, "Les nouveaux diagnostics biologiques", *La Recherche*, May 1987, pp. 658-668.

    R.D. Schmid, "Trends in biosensors", *Biofutur*, March 1988, pp. 37-41.
34. Constantin Bona, "Les vaccins du futur", *La Recherche*, May 1987, pp. 672-682.

35. Wilfred Malenbaum, *World Demand for Raw Materials in 1985 and 2000*, McGraw Hill, New York, 1978.
Eric D. Larson, Marc H. Ross, Robert H. Williams, "Beyond the Era of Materials", *Scientific American*, June 1986, pp. 24-31.
36. Eric D. Larson, M.H. Ross, R. H. Williams, *op. cit.*
37. Vincent Pétiard and Annie Bariaud-Fontanel, "La culture des cellules végétales", *La Recherche*, May 1987, pp. 602-610.
38. *Biotechnology and the Changing Role of Government*, *op cit.*
39. Schumpeter, J.A., *Capitalism, Socialism and Democracy*, George Allen and Unwin, London, 1943 (1976 edition), p. 84.
40. Johnson, H.G., "Technological Change and Comparative Advantage: An Advanced Country's Viewpoint", *Journal of Trade Law*, Vol. 9, 1975.
41. *Impact of New and Emerging Technologies on Trade and Development*, TD/B/C.6/136, Geneva, UNCTAD, 1986.
42. Freeman, C., and Perez, C., "The Diffusion of Technological Innovation and Changes of Techno-economic Paradigms", Paper given at the *Conference on Innovation Diffusion*, Ca Dolfin, Venice, March 1986.
43. Dosi, G., "Technical paradigms and technological trajectories – A suggested interpretation of the determinants and directions of technical change", *Research Policy*, vol. 11, No. 4, 1982.
44. Piore, M.J. and Sabel, C.F., *The Second Industrial Divide: Possibilities of Prosperity*, Basic Books, New York, 1984.
45. Chesnais, F., "Science, Technology and Competitiveness", *STI Review*, No. 1, OECD, Paris, 1986.
46. *Cf. inter alia*, Ohmae, K., *Triad Power: The Coming Shape of Global Competition*, Basic Books, New York, 1985; and Porter, M.E., *Competition in Global Industries*, Harvard Business School Press, Boston, Mass., 1986.
47. *Commercial Biotechnology*, *op. cit.*
48. *Cf. inter alia*, Kaldor, N., "The Role of Increasing Returns, Technical Progress and Cumulative Causation in the Theory of International Trade and Economic Growth", in *Les formes actuelles de la concurrence dans les échanges internationaux*, Colloque ISMEA, Paris, 1980; and Soete, L. and Dosi, G., *Technology Gaps and Cost-Based Adjustments: Some Explorations on the Determinants of International Competitiveness* (paper prepared for the 1983 OECD Workshop on Technological Indicators and the Measurement of Performance in International Trade).
49. *Commercial Biotechnology*, *op. cit.*
50. The longest list is the one discussed in *Commercial Biotechnology*, *op. cit.*
51. Yuan, R.T., *Biotechnology in Western Europe*, mimeo, International Trade Administration, US Department of Commerce Washington, D.C., 1987.
52. Chesnais, F., "Technological Cumulativeness, the Appropriation of Technology and Technological Progressiveness in Concentrated Market Structures", Paper given at the *Conference on Technology Diffusion*, Ca Dolfin, Venice, March 1986.
53. Peters, L.M., *Technical Network Betwwen US and Japanese Industry*, Center for Science and Technology Policy, Rensselaer Polytechnic Institute, Troy, N.Y., 1987.
54. Bustall, M., Dunning, J.H., and Lake A., *Multinational Enterprises Governments and Technology: The Pharmaceutical Industry*, OECD, Paris, 1981.
55. *The Pharmaceutical Industry: Trade Related Issues*, OECD, Paris, 1985.
56. Yuan, R.T., "An Overview of Biotechnological Transfer in our International Context", *Genetic Engineering News*, March 1987.

57. *US Industrial Outlook, 1988*, US Department of Trade, International Trade Administration, Washington, D.C., 1988.

58. Ruivenkamp, G., "The Impact of Biotechnology on International Development: Competition between Sugar and New Sweeteners", in *New Technologies and Third World Development*, Forschungsinstitut der Friedrich-Ebert-Stiftung, Germany, 1986.

59. *Europe Within the World Food System: Biotechnology and New Strategic Options*, Explanatory Dossier 11, FAST II, Bruxelles, (by J. Wilkinson), FAST, 1987.

60. *Impact of New and Emerging Technologies on Trade and Development*, UNCTAD, *op. cit.*

61. *Cf.* Marstrand, P.K., "Production of microbial protein: A study of the development and introduction of a new technology", *Research Policy*, Vol. 10, No. 2, 1981 and Litchfield, J.H., "Single-cell Proteins", *Science*, No. 219, 1983.

62. Kenny, M. and Sasson, A., "Biotechnologies in farming and food systems" in Ann Johnston and Albert Sasson (ed.), *New Technologies and Development*, UNESCO, Paris, 1986.

63. *Impacts of Multinational Enterprises on National Scientific and Technical Capacity: The Food Industry*, OECD, 1979.

64. Kenny, M. and Buttel, F., "Biotechnology: Prospects and Dilemmas for Third World Development", *Development and Change*, Vol. 16, No. 1, June 1985.

65. For a recent review of studies see Brainard, R. and Fullgrable, K., "Technology and Jobs", *STI Review*, No. 1, OECD, Paris, 1986, pp. 9-46.

66. *Biotechnology, International Trends and Perspectives*, by A. Bull, G. Holt, M. Lilly, OECD, Paris, 1982, p. 58.

67. *Biotechnology News*, Vol. 3, No. 22, 15th Nov. 1983, p. 1.

68. *Biotechnology and the Changing Role of Government*, *op. cit.*, Part I, Chapter IV, and Part II, Chapter III, 5.

69. Gerd Junne, "The Impact of Biotechnology on European Agriculture" paper presented at the *International Seminar on Social Dimensions of Biotechnology; Towards a European Policy*, Dublin, 12th November, 1987.

70. A.J. Hacking, *op. cit.*, pp. 63, 94, 117.

71. *Long-Term Economic Impacts of Biotechnology*, unpublished Progress Report, OECD, 1986.

72. *Longterm Economic Impacts of Biotechnology, An International Survey*, conducted by Dr. Rüdiger Hoeren, Prognos AG, Basle, on behalf of the OECD, unpublished report, OECD, 1988.

Annex 1

EXPERTS WHO ATTENDED THE OECD EXPERT SEMINAR ON LONG-TERM ECONOMIC IMPACTS OF BIOTECHNOLOGY SPONSORED BY THE GOVERNMENT OF THE UNITED KINGDOM, 4TH-5TH FEBRUARY 1988, ADMIRALTY HOUSE, WHITEHALL, LONDON

Chairman

Prof. Roger Whittenbury
Chairman of the UK Biotechnology
Advisory Group
University of Warwick
United Kingdom

Dr. Heik Afheldt
Publisher and Managing Editor
Wirtschaftswoche,
Düsseldorf, Germany
Former Managing Director
Prognos AG
Basle, Switzerland

Dr. Mark Cantley
CUBE
Directorate-General
Science, Research and Development
Commission of the European Communities
Brussels, Belgium

Prof. Christopher Freeman
Science Policy Research Unit
University of Sussex
United Kingdom

Dr. Yves M. Galante
Director
MAS Biotec
Milan, Italy

Dr. Riccardo Galli
Professor of Technology Assessment
University of Milan
Italy

Dr. Ralph Hardy
President
Boyce-Thompson Institute
for Plant Research
Cornell University
Deputy Chairman of Biotechnical
International
Ithaca, New York,
United States

Dr. Rüdiger Hoeren
Prognos AG
Basle, Switzerland

Dr. Ernest Jaworski
Director of Biological Sciences
Monsanto Company
St. Louis, Missouri
United States

Professor Dr. Staffan Josephson
Director of Research
Kabi Gen AB, Stockholm
Sweden

Professor Isao Karube
Research Laboratory of Resources
Utilisation,
Tokyo Institute of Technology
Japan

Professor Jakob Nüesch
Director of Pharmaceutical
Research and Biotechnology
Ciba-Geigy AG
Basle, Switzerland

Dr. Lois S. Peters
Senior Research Scientist
School of Management
Rensselaer Polytechnic Institute
Troy, New York, United States

Dr. Brian Richards
Chairman of the Company
British Bio-technology Ltd.
Oxford
United Kingdom

Professor Schilperoort
Professor of Plant Molecular
Biology
Biochemistry Laboratory
University of Leiden
Chairman of the
Advisory Committee on
Biotechnology in the Netherlands

Professor G. Schmidt-Kastner
Director
Biotechnical Process Development
Bayer AG
Wuppertal, Germany

Dr. Hans Herman Schöne
Director
Pharmaceuticals & Biochemical
Research
Hoechst AG, Frankfurt, a.M
Germany

Dr. Margaret Sharp
Science Policy Research Unit
University of Sussex
United Kingdom

Professor Alexander Stavropoulos
VIORYL S.A.
Athens, Greece

Prof. Daniel Thomas
Head of Programme Mobilisateur
University of Compiègne
Research Center
Compiègne, France

UNITED KINGDOM OFFICIALS WHO ATTENDED THE JOINT SESSION ON 5th FEBRUARY 1988

Chief Scientists or Deputies

Dr. Ron F. Coleman
Department of Trade and Industry

Dr. John Rae
Department of Energy

Dr. David Fisk
Department of Environment

Dr. Graham Pearson
Ministry of Defence
Chemical Defense Establishment

Dr. Jeremy S. Metters
Department of Health &
Social Security

Dr. David W.P. Shannon
Ministry of Agriculture,
Fisheries & Foods

Dr. Tom Crossett
Ministry of Agriculture,
Fisheries & Foods

Department of Trade & Industry Officials

Mr. Alex Williams
Government Chemist
Laboratory of the Government Chemist
(LGC)

Dr. Roy Dietz
Deputy Director
LGC

Dr. Peter B. Baker
LGC

Mrs. Chris Hewitt
LGC

Miss Alison J. Mo
LGC

Miss Jenny Gray
LGC

Dr. Robin H. Aram
Research & Technology Policy Division

Mr. John M. Barbet
Economics Division

Mr. Spencer Nathan
Economics Division

Other Government Officials

Mrs. Elisabeth Attridge
Ministry of Agriculture
Fisheries & Food

Mr. Brian Ager
Health & Safety Executive

Mr. Richard Clifton
Health & Safety Executive

OECD OFFICIALS

Dr. John Bell
Head, Science and Technology
Policy Division

Dr. Salomon Wald
Biotechnology Unit

Miss Bruna Teso
Biotechnology Unit

Dr. François Chesnais

Annex 2

COMPANIES WHICH HAVE BEEN INTERVIEWED
(See Chapter II)

AJINOMOTO CO. INC.	Japan
AKZO	Netherlands
ANHEUSER BUSCH	United States
ALLELIX	Canada
ANTIBIOTICOS SA	Spain
ARTHUR GUINNESS SON & CO	Ireland
ASAHI CHEMICAL	Japan
AYUSO SA	Spain
BEATRIX FOOD	United States
BECK UND CO	Germany
BIO MEGA	Canada
BIOTECHNICA	United Kingdom
BSN	France
CARL-BIOTECH	Denmark
CDC LIFE SCIENCES	Canada
CETUS	United States
CHIRON	United States
CIBA-GEIGY	Switzerland
CIPAN	Germany
CLONATECH	France
COMPAGNIE GENERALE DES EAUX	France
CONNAUGHT LABORATORIES	Canada
DAMON BIOTECH	United States
DE DANSKE SUKKERFABRIKKER	Denmark
DSM	Netherlands
EISAI CO LTD	Japan
EUMIG FOHNSDORF	Austria
FERMIGN	Germany
FINNISH SUGAR CO LTD	Finland
GENERALE BISCUIT	Finland
GIST BROCADES	Netherlands
THE GREEN CROSS CORP	Japan
HENKEL	Germany
HEINZ	United States
HUTTON	United States
IAF BIOCHEM	Canada
IBM	United States
ICI	United Kingdom
IMMUNO DIAG	Austria
IMMUNOFARMA	Portugal

INGENASA	Spain
INSTITUTO GULBENKIAN DE CIENCIA	Portugal
JOHN LABATT	Canada
F. HOFFMANN-LA-ROCHE	Switzerland
KABIGEN	Sweden
KELLOGG	United States
KEMIRA OY	Finland
KIRIN	Japan
KODAK	United States
KYOWA HAKKO	Japan
LIMAGRAIN	France
INSTITUT MERIEUX	France
MITSUBISHI CHEMICAL	Japan
MONSANTO	United States
MONTEDISON/FARMITALIA	Italy
NAKANO VINEGAR	Japan
NISSHO IWAI	Japan
NOC TECH	Ireland
NORDISK INSULIN LABORATORIUM	Denmark
NOVO INDUSTRIES	Denmark
ORION CORPORATION	Finland
ORSAN	Finland
PFEIFER UND LANGEN	Denmark
PHARMACIA	Sweden
PHILLIPS-PROVESTA	United States
PROGEN	Germany
PULP AND PAPER RESEARCH INSTITUTE OF CANADA	Canada
RECORDATI	Italy
RHM	United Kingdom
RHONE POULENC	France
PRODUCTORS RIBA	Spain
RINTEKO OY	Finland
SANDOZ	Switzerland
SCA	Sweden
SCHWEPPES-CADBURY	United Kingdom
SEARLE	United Kingdom
SOCIETA PRODUZIONE ANTIBIOTICI	Italy
SOLVAY	Belgium
SUMITOMO	Japan
SUNTORY	Japan
TAKEDA	Japan
TEIJIN	Japan
THOMAE	Germany
TIENSE SUIKER	Belgium
TORAY	Japan
TOYO JOZO	Japan
TULLNER ZUCKERFABRIK	Austria
UNILEVER	Netherlands
VOGELBUSCH	Austria
UPJOHN	United States
WARCOING	Belgium
WELLCOME	United Kingdom
WEYERHAEUSER	United States
YOMO-CENTRO SPERIMENTALE DEL LATTE	Italy

WHERE TO OBTAIN OECD PUBLICATIONS
OÙ OBTENIR LES PUBLICATIONS DE L'OCDE

ARGENTINA - ARGENTINE
Carlos Hirsch S.R.L.,
Florida 165, 4° Piso,
(Galeria Guemes) 1333 Buenos Aires
Tel. 33.1787.2391 y 30.7122

AUSTRALIA - AUSTRALIE
D.A. Book (Aust.) Pty. Ltd.
11-13 Station Street (P.O. Box 163)
Mitcham, Vic. 3132 Tel. (03) 873 4411

AUSTRIA - AUTRICHE
OECD Publications and Information Centre,
4 Simrockstrasse,
5300 Bonn (Germany) Tel. (0228) 21.60.45
Gerold & Co., Graben 31, Wien 1 Tel. 52.22.35

BELGIUM - BELGIQUE
Jean de Lannoy,
Avenue du Roi 202
B-1060 Bruxelles Tel. (02) 538.51.69

CANADA
Renouf Publishing Company Ltd
1294 Algoma Road, Ottawa, Ont. K1B 3W8
Tel: (613) 741-4333
Stores:
61 rue Sparks St., Ottawa, Ont. K1P 5R1
Tel: (613) 238-8985
211 rue Yonge St., Toronto, Ont. M5B 1M4
Tel: (416) 363-3171
Federal Publications Inc.,
301-303 King St. W.,
Toronto, Ont. M5V 1J5 Tel. (416)581-1552
Les Éditions la Liberté inc.,
3020 Chemin Sainte-Foy,
Sainte-Foy, P.Q. G1X 3V6, Tel. (418)658-3763

DENMARK - DANEMARK
Munksgaard Export and Subscription Service
35, Nørre Søgade, DK-1370 København K
Tel. +45.1.12.85.70

FINLAND - FINLANDE
Akateeminen Kirjakauppa,
Keskuskatu 1, 00100 Helsinki 10 Tel. 0.12141

FRANCE
OCDE/OECD
Mail Orders/Commandes par correspondance :
2, rue André-Pascal,
75775 Paris Cedex 16 Tel. (1) 45.24.82.00
Bookshop/Librairie : 33, rue Octave-Feuillet
75016 Paris
Tel. (1) 45.24.81.67 or/ou (1) 45.24.81.81
Librairie de l'Université,
12*a*, rue Nazareth,
13602 Aix-en-Provence Tel. 42.26.18.08

GERMANY - ALLEMAGNE
OECD Publications and Information Centre,
4 Simrockstrasse,
5300 Bonn Tel. (0228) 21.60.45

GREECE - GRÈCE
Librairie Kauffmann,
28, rue du Stade, 105 64 Athens Tel. 322.21.60

HONG KONG
Government Information Services,
Publications (Sales) Office,
Information Services Department
No. 1, Battery Path, Central

ICELAND - ISLANDE
Snæbjörn Jónsson & Co., h.f.,
Hafnarstræti 4 & 9,
P.O.B. 1131 – Reykjavik
Tel. 13133/14281/11936

INDIA - INDE
Oxford Book and Stationery Co.,
Scindia House, New Delhi 110001
Tel. 331.5896/5308
17 Park St., Calcutta 700016 Tel. 240832

INDONESIA - INDONÉSIE
Pdii-Lipi, P.O. Box 3065/JKT.Jakarta
Tel. 583467

IRELAND - IRLANDE
TDC Publishers - Library Suppliers,
12 North Frederick Street, Dublin 1
Tel. 744835-749677

ITALY - ITALIE
Libreria Commissionaria Sansoni,
Via Benedetto Fortini 120/10,
Casella Post. 552
50125 Firenze Tel. 055/645415
Via Bartolini 29, 20155 Milano Tel. 365083
La diffusione delle pubblicazioni OCSE viene assicurata dalle principali librerie ed anche da :
Editrice e Libreria Herder,
Piazza Montecitorio 120, 00186 Roma
Tel. 6794628
Libreria Hœpli,
Via Hœpli 5, 20121 Milano Tel. 865446
Libreria Scientifica
Dott. Lucio de Biasio "Aeiou"
Via Meravigli 16, 20123 Milano Tel. 807679

JAPAN - JAPON
OECD Publications and Information Centre,
Landic Akasaka Bldg., 2-3-4 Akasaka,
Minato-ku, Tokyo 107 Tel. 586.2016

KOREA - CORÉE
Kyobo Book Centre Co. Ltd.
P.O.Box: Kwang Hwa Moon 1658,
Seoul Tel. (REP) 730.78.91

LEBANON - LIBAN
Documenta Scientifica/Redico,
Edison Building, Bliss St.,
P.O.B. 5641, Beirut Tel. 354429-344425

MALAYSIA/SINGAPORE - MALAISIE/SINGAPOUR
University of Malaya Co-operative Bookshop Ltd.,
7 Lrg 51A/227A, Petaling Jaya
Malaysia Tel. 7565000/7565425
Information Publications Pte Ltd
Pei-Fu Industrial Building,
24 New Industrial Road No. 02-06
Singapore 1953 Tel. 2831786, 2831798

NETHERLANDS - PAYS-BAS
SDU Uitgeverij
Christoffel Plantijnstraat 2
Postbus 20014
2500 EA's-Gravenhage Tel. 070-789911
Voor bestellingen: Tel. 070-789880

NEW ZEALAND - NOUVELLE-ZÉLANDE
Government Printing Office Bookshops:
Auckland: Retail Bookshop, 25 Rutland Stseet,
Mail Orders, 85 Beach Road
Private Bag C.P.O.
Hamilton: Retail: Ward Street,
Mail Orders, P.O. Box 857
Wellington: Retail, Mulgrave Street, (Head Office)
Cubacade World Trade Centre,
Mail Orders, Private Bag
Christchurch: Retail, 159 Hereford Street,
Mail Orders, Private Bag
Dunedin: Retail, Princes Street,
Mail Orders, P.O. Box 1104

NORWAY - NORVÈGE
Narvesen Info Center – NIC,
Bertrand Narvesens vei 2,
P.O.B. 6125 Etterstad, 0602 Oslo 6
Tel. (02) 67.83.10, (02) 68.40.20

PAKISTAN
Mirza Book Agency
65 Shahrah Quaid-E-Azam, Lahore 3 Tel. 66839

PHILIPPINES
I.J. Sagun Enterprises, Inc.
P.O. Box 4322 CPO Manila
Tel. 695-1946, 922-9495

PORTUGAL
Livraria Portugal, Rua do Carmo 70-74,
1117 Lisboa Codex Tel. 360582/3

SINGAPORE/MALAYSIA - SINGAPOUR/MALAISIE
See "Malaysia/Singapor". Voir «Malaisie/Singapour»

SPAIN - ESPAGNE
Mundi-Prensa Libros, S.A.,
Castelló 37, Apartado 1223, Madrid-28001
Tel. 431.33.99
Libreria Bosch, Ronda Universidad 11,
Barcelona 7 Tel. 317.53.08/317.53.58

SWEDEN - SUÈDE
AB CE Fritzes Kungl. Hovbokhandel,
Box 16356, S 103 27 STH,
Regeringsgatan 12,
DS Stockholm Tel. (08) 23.89.00
Subscription Agency/Abonnements:
Wennergren-Williams AB,
Box 30004, S104 25 Stockholm Tel. (08)54.12.00

SWITZERLAND - SUISSE
OECD Publications and Information Centre,
4 Simrockstrasse,
5300 Bonn (Germany) Tel. (0228) 21.60.45
Librairie Payot,
6 rue Grenus, 1211 Genève 11
Tel. (022) 31.89.50
Maditec S.A.
Ch. des Palettes 4
1020 – Renens/Lausanne Tel. (021) 635.08.65
United Nations Bookshop/Librairie des Nations-Unies
Palais des Nations, 1211 – Geneva 10
Tel. 022-34-60-11 (ext. 48 72)

TAIWAN - FORMOSE
Good Faith Worldwide Int'l Co., Ltd.
9th floor, No. 118, Sec.2, Chung Hsiao E. Road
Taipei Tel. 391.7396/391.7397

THAILAND - THAILANDE
Suksit Siam Co., Ltd., 1715 Rama IV Rd.,
Samyam Bangkok 5 Tel. 2511630
INDEX Book Promotion & Service Ltd.
59/6 Soi Lang Suan, Ploenchit Road
Patjumamwan, Bangkok 10500
Tel. 250-1919, 252-1066

TURKEY - TURQUIE
Kültur Yayinlari Is-Türk Ltd. Sti.
Atatürk Bulvari No: 191/Kat. 21
Kavaklidere/Ankara Tel. 25.07.60
Dolmabahce Cad. No: 29
Besiktas/Istanbul Tel. 160.71.88

UNITED KINGDOM - ROYAUME-UNI
H.M. Stationery Office,
Postal orders only: (01)873-8483
P.O.B. 276, London SW8 5DT
Telephone orders: (01) 873-9090, or
Personal callers:
49 High Holborn, London WC1V 6HB
Branches at: Belfast, Birmingham, Bristol, Edinburgh, Manchester

UNITED STATES - ÉTATS-UNIS
OECD Publications and Information Centre,
2001 L Street, N.W., Suite 700,
Washington, D.C. 20036 - 4095
Tel. (202) 785.6323

VENEZUELA
Libreria del Este,
Avda F. Miranda 52, Aptdo. 60337,
Edificio Galipan, Caracas 106
Tel. 951.17.05/951.23.07/951.12.97

YUGOSLAVIA - YOUGOSLAVIE
Jugoslovenska Knjiga, Knez Mihajlova 2,
P.O.B. 36, Beograd Tel. 621.992

Orders and inquiries from countries where Distributors have not yet been appointed should be sent to:
OECD, Publications Service, 2, rue André-Pascal, 75775 PARIS CEDEX 16.

Les commandes provenant de pays où l'OCDE n'a pas encore désigné de distributeur doivent être adressées à :
OCDE, Service des Publications. 2, rue André-Pascal, 75775 PARIS CEDEX 16.

72380-1-1989

OECD PUBLICATIONS, 2, rue André-Pascal, 75775 PARIS CEDEX 16 - No. 44663 1989
PRINTED IN FRANCE
(93 89 01 1) ISBN 92-64-13196-5